T0314239

Technological Change and the Evolution of Corporate Innovation

PREST/CRIC STUDIES IN SCIENCE, TECHNOLOGY AND INNOVATION

General Editors: Luke Georghiou, *Director, PREST* and Stan Metcalfe, *Stanley Jevons Professor of Economics and ESRC Centre for Research on Innovation and Competition, University of Manchester, UK*

This series is a forum for the outstanding research from Manchester's two leading social science research institutes – PREST (Policy Research in Engineering, Science and Technology) and CRIC (the ESRC Centre for Research on Innovation and Competition).

The books in this series reflect their concern with the economics, social and managerial implications of science and innovation. Emphasis is given to science and technology policy and the role that innovation plays in competitiveness, particularly in the new knowledge-intensive global economy.

This interdisciplinary series will include some of the best theoretical and empirical work in the field, and will be invaluable to scholars, students and policy makers.

Titles in the series include:

Knowledge and Innovation in the New Service Economy
Edited by Birgitte Andersen, Jeremy Howells, Richard Hull, Ian Miles and Joanne Roberts

Technological Change and the Evolution of Corporate Innovation
The Structure of Patenting 1890–1990
Birgitte Andersen

Technological Change and the Evolution of Corporate Innovation

The Structure of Patenting 1890–1990

Birgitte Andersen

Senior Lecturer, Birkbeck College, University of London, UK

Honorary Associate Fellow, ESRC Centre for Research on Innovation and Competition (CRIC), University of Manchester, UK

PREST/CRIC STUDIES IN SCIENCE, TECHNOLOGY AND INNOVATION

Edward Elgar

Cheltenham, UK • Northampton, MA, USA

Published by
Edward Elgar Publishing Limited
Glensanda House
Montpellier Parade
Cheltenham
Glos GL50 1UA
UK

Edward Elgar Publishing, Inc.
136 West Street
Suite 202
Northampton
Massachusetts 01060
USA

A catalogue record for this book
is available from the British Library

Library of Congress Cataloguing in Publication Data

Andersen, Birgitte, 1967–
Technological change and the evolution of the corporate innovation : the structure of patenting 1890–1990 / Birgitte Andersen.
p. cm. — (PREST/CRIC studies in science, technology and innovation)
Includes bibliographical references and index.
1. Technological innovations. 2. Patents—History. I. Title. II. Series.

T173.8 .A49 2001
658.4'063—dc21

00–047643

ISBN 1 84064 121 5

Printed and bound in Great Britain by MPG Books Ltd, Bodmin, Cornwall

Contents

Preface

My interest in the field of technological change and corporate innovation dates back from when I started my PhD in the Department of Economics at the University of Reading. Special thanks are due to Professor John Cantwell who introduced me to the world of patent statistics, which I have used extensively in this book, and for collaboration and inspiration. I would also like to thank him for sharing the patent database he has developed with the assistance of the US Patent and Trademark Office.

When I finished my PhD and became an ESRC Research Fellow at the Centre for Research on Innovation and Competition (CRIC) at the University of Manchester and UMIST, I was able to broaden my knowledge in this field. Specific thanks go to Professor Stanley Metcalfe for discussions and profound comments. Thanks also go to Professor Rod Coombs, Professor Denise Osborn, Dr Vivien Walsh, Professor Ian Miles, and Dr Sally Randles for helpful guidance and criticism on various parts of this book. In addition, I would like to thank CRIC's Administrator Ms Sharon Hammond for providing a research environment with such an illuminating spirit. Thanks also to Ms Deborah Woodman and Mrs Sharon Dalton for splendid typographical assistance.

I would also like to thank Professor Nick von Tunzelman, Dr Felicia Fai, Professor Keith Pavitt and Dr Pari Patel for discussions and comments. Acknowledgement for comments on pieces of my work informing this book also go to Professor Klaus Nielsen, Professor Bo Carlsson, Professor Bengt-Åke Lundvall, Professor Gunner Eliasson, Professor James Utterback, and Professor Francisco Louçã.

Moreover, I wish to acknowledge financial support from the Danish Social Science Research Council (SSF) and the Danish Research Academy which facilitated my initial research in this field, as well as acknowledging the Economic and Social Research Council (ESRC) of the UK for my Research Fellowship, until I became a Senior Lecturer at Birkbeck College, University of London.

Finally, thanks to all the staff at Edward Elgar, especially Ms Dymphna Evans, for a pleasant partnership.

Birgitte Andersen

1. Introduction

1.1. PROBLEM SETTING AND AIM OF BOOK

We know that the creation and diffusion of technological knowledge is the heart of modern economic growth (Kuznets 1930a, Schumpeter 1939), and that economic growth and investment-demand stimulate technological development (Schmookler 1966). It is also widely accepted that technological change and economic growth is channelled in certain directions rather than others (Rosenberg 1976, Nelson and Winter 1977, 1982), and that 'systems' can act as stimuli or constraints in the rate and direction of this process (Freeman 1987, Lundvall 1992, Carlsson and Stankiewicz 1991). Such insights have mainly been centred around a notion of structural change. They have helped to shed light on the specificity of national institutions and how technology is differentiated across firms, industries, regions, technological sectors and over time, which all have implications on the distribution of global resources and wealth.

However, the empirical work has been rather comparative-static, and descriptive, although dynamic in its theoretical reasoning. Only very limited empirical effort has been devoted to understanding the *processes* by which technological structures, and the innovative and competitive landscapes, emerge, transform, grow, decline and demise. This in its turn has left the theoretical reasoning mainly based upon 'beliefs' and simulation studies, with insufficient empirical evidence. Equally, the insufficient empirical evidence has also hampered a theoretical refinement of the processes that prevail within evolutionary economics, which still seems somewhat approximate or vague in their generality. Hence, our theoretical as well as empirical knowledge of change and development processes is still seriously incomplete.

The aim of this volume is simple. It is to explore explicitly some central dynamic processes of change and development within evolutionary economics. It is mainly about understanding open-ended transformation processes, as well as understanding the underpinning 'rules'. The notion of 'rules', in this context, is used in a way resembling 'rules of the game' in North (1990).

The book challenges many of the 'beliefs' within Schumpeterian reasoning with respect to the dynamic elements of technological change and corporate innovation, by bringing empirical rigour to their story. In this context, it brings detail and specificity to the broad nature of Schumpeterian reasoning. By specificity is meant the historical findings about the instituted nature of the 'rules' underpinning change and development processes.

The book reveals and interprets the processes by which instituted technological change, governed by major technological epochs or regimes of waves of development, have become more interrelated across technological fields, and how the path-dependent evolution of corporate innovation have become increasingly complex in the knowledge bases upon which it draws. The book also reveals how different rates of development (i.e. catching up and accumulation) of technologies and firms creates (or selects) new technological structures with new industrial scopes. This, in its turn, transforms the innovative and competitive landscapes, in an open-ended process of continuous change and development.

Overall, by disclosing the dynamic processes and the underpinning 'rules' of change and development historically, this book has implications for the interpretation of history, the evolution of industrialised societies, and the social sciences.

An explicit recognition of processes of path-dependency and cumulativeness, as well as processes of development, change and emergence, makes this book a potential supporter of historical studies.

1.1.1. The Empirical Basis

The exploitation of knowledge embodied in product and process innovations, new ideas, or related to intangible assets and symbolic material, is in most mature economies protected through the use of Intellectual Property Rights (IPRs). The empirical research in this book is based upon a US patent database comprising all individual and corporate (US national and international) patents, granted in the US during the period from 1890 to 1990, and classified at a detailed level of disaggregation.

The rise of corporate capitalism during the nineteenth and (especially early) twentieth centuries, based on the manufacturing system and associated with physical artefacts and production processes, helped to shape and push the development and structure of the modern patent system with its emphasis on protecting physical artefacts centred on new products and production processes (Noble 1979, Sullivan 1990, Andersen and Howells 2000). The macro-economic efficiency in the governance of IPR, and the microeconomic aspects of such IPR systems, economists have taken advantage of. Much economic research, especially with respect to patents and the manufacturing system, has been on the possibilities of using them as indicators and specialisation indices (as is the case of this book), or related to their economic value or worth.[1]

Hence, using patents granted in the US, historically, all the chapters in this book develop and/or apply different indicators and specialisation indices, as well as different quantitative and statistical methods of various degrees of sophistication. It is primarily based upon patent statistics combined with panel data analysis. The main advantage of using US patent data for the purpose of this book is that they are recorded historically and at detailed level of disaggregation, which makes it possible to analyse trajectories and changes in structures at a very detailed level.

Since the analysis and results of this book rest mainly on what information patent data can provide, a discussion of the possibilities and problems, as well as appropriate use of such measures is carried out in Chapter 2. Chapter 2 also presents the organisation of the US patent database on which the book is based.

In studying time series of US patenting activity, this book can best be considered as being in line with the empirical work of Kuznets (1930a) and Schmookler (1966) and their work on US patenting activity, which they addressed in relation to production and prices as well as sectoral investment demand. However, instead of researching the 'causes' and 'effects' concerning why innovation and investment and other economic variables appear (discussions raised by Kuznets and Schmookler), this book focuses on the processes by which development and changes in industrialised societies take place.

1.2. A PREVIEW OF THE BOOK

Seven major interrelated themes will provide the basic frameworks by which the dynamic processes, and the 'rules' underpinning such, are studied in this book. The themes will now be presented together with a preview of the chapters of the book. Of course the seven themes are much broader than any of the chapters addressed within them, but here it is found profitable to present the themes around which the chapters are based (rather than the other way around) to provide an overview of the application of the book. Hence, the chapters serve as evidence (or case study accounts) for the dynamic processes which prevail within the themes. An outlook on the dynamic processes and underpinning 'rules' as the unit of analysis is subsequently carried out in Section 1.3.

[1] Still, the study of IPRs covers a diverse range of subjects, disciplines and legal regimes. As such it includes a whole set of different types of legal statute such as property, contract and competition law as well as involving a wide spectrum of economic and social issues relating to, for example, trade, monopoly and competition issues.

Theme 1: The evolution of technological paradigms governing trajectories of opportunities

The first theme addressed within the book examines the evolution of technological opportunities. It is the overall nature of such which define the opportunities for further innovations and some basic procedures used to exploit them, and this has been referred to as technological paradigms by institutional evolutionary economists (Freeman and Perez 1988). Hence it is the characteristics of such paradigms which sets the overall directions of development, whereas the specific paths of individual technological sectors have been defined as technological trajectories (Dosi 1982).

Supposing that evolution is cumulative, incremental and path-dependent by nature (see Section 1.3), and that technological change is channelled along certain paths of development rather than others, Chapter 3 derives the nature by which path-dependent processes have created structural changes in the accumulation and catching up of technological opportunities historically. 'Micro to macro' analysis (see Section 1.3) investigating cross-sectional interaction of technological fields reveals how technological systems build upon knowledge embodied in old ones rather than substitute it. In this context, Chapter 3 also shows how techno-economic systems, governed by overall technological paradigms or regimes, have become more interrelated across different technological fields. The extent to which evolutionary processes of technological advance have been 'instituted' along certain paths or trajectories of development is also revealed.

Theme 2: The S-shaped image of the technological growth cycle and the diversity of technology dynamics

Since the works of the business cycle theorists in the 1930s, no attempts have been made to study empirically the long-term evolution paths of individual technologies starting with long time series.

Hence, the second theme of the book (Chapter 4) considers the cyclical nature of technological development, in which special attention is paid to the presence of the Sigmoid-shaped (or *S*-shaped) image of the technological growth cycle. The underlying processes by which such cycles can be identified is elaborated upon theoretically and empirically, in relation to deriving and identifying innovation trajectories (or growth paths) of related technological fields. The biological growth model in logistic form as an appropriate approximation to describe the evolution or cyclical growth paths is revealed.

Drawing upon the work of business-cycle economists (especially Schumpeter 1939 and Kuznets 1940), the nature of the technological growth curve will be conceptualised and extrapolated in depth. Each technological growth cycle will be characterised in relation to the parameters and properties which guide its evolutionary path. Special attention is given to characterising Schumpeter's four phases of cyclical development (prosperity, recession, depression and revival), and to identifying the duration of each technological growth cycle as well as the timing of takeoffs in innovative activity. Furthermore, the possibility of combined or synchronised technological takeoff (which are subsequently associated with waves in technological development, see Chapter 5) will be investigated.

The results have implications for our understanding of the evolutionary paths of individual technologies. Moreover, by comparing the evolution paths and cycle properties of the identified innovation trajectories, the diversity of such will also be reviled. Such diversity have implications for the way we might understand technological systems, and evolving waves of innovation, change and development.

Theme 3: Long waves as a phenomenon in real quantities

The third theme of the book moves the analysis to consider the processes of secular moments in technological time series. As argued by Rosenberg and Frischtak (1984), Kondratiev's (1925) hypothesis of the existence of long waves in economic life has given rise to two distinct lines of historical research: one centred around price (or interest rate) cycles and the other focused on long waves as a phenomenon in real quantities. It is the latter type of wave which will be concentrated on in Chapter 5.

In a Schumpeterian vein, special attention is given to analysing and deriving the possibility of wave-like patterns in clusters of 'established innovation trajectories', as opposed to 'basic' or radical innovations (Chapter 5). That is, whereas most empirical studies on long waves as a phenomenon in real quantities mainly have been around clustering around the *appearance* of 'basic' or radical innovations, this chapter instead focuses on Freeman, Clark and Soete (1982), who argued that it is not the *appearance* of basic (unrelated) innovations so much as their *diffusion*, which creates the clustering effect of related product, process, and applications innovations.

Thus, Chapter 5 is concerned with the type of innovation wave related to solving of techno-economic problems with respect to the development of innovation trajectories. Such views have also been related to Schmookleran bandwagon effects (Schmookler 1966), or associated with the Schumpeterian 'entrepreneurial swarming' processes (Schumpeter 1939). Freeman and Perez (1988) relate innovations clusters with the emergence of techno-economic paradigms.

In the context of Chapter 5, the trajectories of technological evolution (identified in Chapter 4) are linked with 'system thinking', where each identified wave of innovation trajectories is treated in a systemic context, as a changing population of innovation trajectories with changing internal structures (see Section 1.3 on 'The system perspective'). The empirical results are discussed in relation to an interpretation of Kondratiev's (1925) notion of the third, fourth and fifth long wave.

Theme 4: Industrial Dynamics

The system perspective (developed within theme 3 above) is then applied more or expanded in the fourth theme of this book which has its origin within Industrial Dynamics. Industrial Dynamics is a 'problem oriented' approach to a particular set of questions, rather than a subfield within a discipline. The three principle questions are related to investment in productive capacity, changes in technique, and product innovation. (For an overview on Industrial Dynamics, see Andersen and Lundvall, forthcoming.)

Within Industrial Dynamics, transformations in industrial structures are caused (or naturally selected) by competitive and cooperative processes in the macro environment or the market place, as well as by the interaction between such processes in the development of corporate capabilities of industries (i.e. the endowments and resource-bases that are developed internally within firms, or distributed in cooperative arrangements among firms and other institutions). Focus is especially on firms' (or groups of firms') changing endowments and resource-bases, driven by changing technology, organisation, and relative capabilities (embodied in routines, habits and practices inherited from the past). Studies have especially been on competitive and cooperative processes, with respect to how firms create and respond (i.e. the behaviour of firms) to innovation. Selection or competition is not constrained to product markets. In the Industrial Dynamics approach long-term competitiveness is as much located within the firm's dynamic supply capabilities (e.g. technological capabilities, adjustment capabilities, local networks), or distributed in cooperative arrangements among firms and other institutions. Hence, Industrial Dynamics is compatible with the resource-based theories of the firm (Penrose 1959, Richardson 1972, Foss 1993, Eliasson 1997). In the management literature similar issues have been addressed with respect to learning organisations. In this framework, firms only optimise locally, if at all, (i.e. there is no generally applicable best corporate practice), since they are constrained in the choices they make and by the resources and capabilities that they control (everything is context specific). Hence, the mainstream economics' production function can not be taken for granted.

One of the most comprehensive attempts to investigate the systemic aspects of innovation with Industrial Dynamics at the micro-economic level is provided by Carlsson and colleagues (1987, 1995, 1997). A central feature of their studies is the emphasis on 'technological systems' (see Chapter 6). Among the several important contributions made is the significance of the (unequal) distribution of what they term economic competence, and the ability to develop and exploit new business opportunities. The Industrial Dynamics literature also points out that economic competence within a region, industrial sector or firm is often built on the integration of very diverse technologies, and how systems can act as barriers as well as stimuli to knowledge accumulation. They also make the important step of studying the systems supporting the development of generic technologies as they evolve through several decades.

Distributed innovation processes within industrial systems supporting the development of generic technologies, as they evolve through several decades, are derived in Chapter 6, which adopt a technological system perspective. The co-evolution between (i) the wave-like development patterns of clusters of innovation trajectories (derived in Chapter 5) and (ii) the corporate capabilities within industrial structures will be identified. Chapter 6 illustrates how the knowledge bases upon which firms draw, exploit and develop have become increasingly complex, and how this co-evolves with the changing business opportunities governed by evolving technological paradigms.

The research within this theme opens up for a discussion of the limitations the standard industrial classification scheme oblige, if we wish to understand firms in relation to the changing technological and competitive landscape in which they operate, and why a system perspective may be more appropriate. Hence, whereas the product cycle literature (Kuznets 1930a, Vernon 1966, Utterback 1996, Utterback and Suarez 1993) has helped us to understand changes in the characteristics of markets, technology and products along associated life cycles within specific industries, the Industrial Dynamics literature typically involves a more interactive and multi-sectoral focus on change, in order to identify the industrial dynamic structures at a given time.

Theme 5: The variety of firms' technological profiles and how it matters for their trajectories

Chapter 6 within theme 4 on Industrial Dynamics helps us to understand how competitive processes (or competition) between firms built technological systems of innovation, in relation to changing business environments historically. We now move on to focus on the strategies and constraints prevailing within individual firms, which are part of such processes.

Reasoning along lines similar to those developed by business scholars and evolutionary economists, it is argued that not all firms are equally competent in adapting to new technological opportunities, as firms differ with respect to their dynamic technological capabilities. That is, they are not equally competent at generating, absorbing and exploring new ideas and technological opportunities. It is here argued that this is due to the localised nature of technological development which creates a variety in the competencies of firms, which again is influenced by their past internal collective learning processes, inherited routines and expertise; and that economic theory needs to be able to identify and classify more specifically how firms differ, in order to explore how the variety of firms is affected by broader technological evolution.

Hence, the fifth theme of the book (Chapter 7) considers the variety of firms in relation to their types of technological competence. Since changes in technological expertise and routines only occur gradually and within certain boundaries over time, special attention is on the dynamic impact the variety of firms' types of technological competencies have on their subsequent path-dependent corporate trajectories. Hence, at a very detailed level, the evolution paths or trajectories of individual firms are derived. The extent to which firms become locked into some specific paths of development, as well as how they change the composition of their competencies within and across changing technological regimes are revealed.

The variety of firms' types of technological competence is developed around a taxonomy, where firms are grouped with respect to their degree of technological leadership, their specialisation in areas that represent their industry's core fields of technological development, and their dispersion or breadth of competence, which is related to their technological size. The path-dependent evolution paths, or trajectories, of firms is analysed and confirmed, with respect to the firms' probability of transition over time.

Hence, identifying the variety of firms' technological profiles and how it matters for their subsequent trajectories, Chapter 7 certainly demonstrates that, while there are limits to the scope of managerial discretion over corporate technological strategy given the specificity of industry paths (as argued by Patel and Pavitt 1998), within the limits of these constraints there is still room for different types of firms and different approaches to types of technological specialisation or business models.

Theme 6: Diffusion of corporate technological leadership in the course of technological growth

The sixth theme of the book considers Schumpeter's specific contribution to business cycle theory (Schumpeter 1939), which deals with the association between cyclical movements in the rate of innovations and the diffusion (or changing distribution) of entrepreneurial ability. It concerns the diffusion processes of corporate technological leadership in the course of technological growth. Referring to Schumpeter's association between cyclical fluctuations in the rate of innovations (as identified in Chapter 4) and the distribution of entrepreneurial ability, it is now commonly argued among diffusion-economists that the overall growth of technological activity is linked to dispersion of corporate technological leadership, or linked to 'swarming' in technological activity using Schumpeter's terminology.

Hence, Chapter 8 examines the dispersion or diffusion of corporate technological leadership in sectors with high technological opportunities historically. The question becomes the extent to which technological leadership, defined as a relatively high specialisation in the areas of greatest technological opportunities, dissipates in the course of technological development. The extent to which existing leaders incrementally consolidate their lead as new technological fields emerge, or at least preserve rather than lose their position, is tested.

Both the degree of catching up in technological leadership in the cause of growth, as well as the moderate extent of historical shifts in the actual composition of corporate technological leadership, is revealed. Thus, path-dependent nature of the trajectories of firms (as also identified in Chapter 7) are also central to the nature by which such diffusion processes form.

Theme 7: The overall nature of open-ended transformation processes

The final theme of this book (Chapter 9) brings together what we have learned in each chapter, in order to discuss and evaluate the results in a broader setting. Special attention is on identifying the stylised facts concerning technological change and the evolution of corporate innovation 1890-1990. It is argued that the nature of structural changes and 'waves' of development is associated with processes (for development, selection and variety creation) which are of more *general* kind, and that this general nature needs to be recognised separately from the inherited and *specific* nature of paradigms and trajectories. It is the latter which provides a framework for understanding the 'rules' which is guiding the characteristics underpinning the change and development processes, which then become instituted.

1.3. BUILDING BLOCKS FOR DYNAMIC PROCESSES

The seven themes represent some of the most dominant and conventional theoretical, empirical and conceptual frameworks by which many of the dynamic processes underlying evolutionary economics are understood and studied. Accordingly, an outlook on the dynamic processes which prevail within those themes will now be presented, as they form the conceptual and empirical basis for how dynamic processes are studied within this book. This will also provide an understanding of the analytical origin of this book.

There can be little doubt that evolutionary economics has been successful in challenging mainstream economics theoretically as well as conceptually, and it has opened a very interesting research agenda. Building upon Nelson and Winter's (1982) critique of mainstream economic models, it is now commonly recognised that mainstream economic models are less than adequate in dealing with the growing impact of technology on the dynamic, changing and unstable economy, as well as less than adequate in dealing with the complexity of firms. What has been referred to is the mainstream economists' awkward treatment of technology which shapes their theories with a static underpinning in their theorising, and that firms are assumed to be homogenous units which act completely rationally. Hence, the problem with the mainstream economics models is that they are far removed from reality. Therefore, a more dynamic approach has been explored in which firms differ and technology is integrated as an endogenous factor in economic theory. In this context, the role of innovation in interactive development and selection processes (i.e. the structural dimensions of the transformations of entities) are central areas examined within evolutionary economics.

When exploring the dynamic processes of technological change and the evolution of corporate innovation, evolutionary economics has provided us with some building blocks which now appear as almost 'beliefs' in economic reasoning. These 'beliefs' are not taken for granted in this book, but they are central to the analytical construction of how the dynamic processes of development and change are explored. Obviously the different building blocks presented below are related (or perhaps they even overlap), but they are for analytical purposes kept separate.

Historical path dependency and instituted processes of change

Rosenberg (1976) and Nelson and Winter (1977, 1982) introduced some perspectives on the nature of technological change and the behaviour of firms and industrialised societies, in relation to economic conceptualisation and long-term growth processes. Emphasis was on how technologies, organisations and the dynamic capabilities (developed from past learning processes and inherent routines) of firms change only incrementally and are cumulated along specific paths or trajectories. Nelson (1991a, 1991b) reviewed the notion of path-dependency in relation to how the boundaries to firms' (or groups of firms') behaviour explain why they differ in their knowledge base, in their capability to innovate, and in their ability to appropriate from the innovations they make.

Shortly after the pioneering works of Rosenberg and Nelson and Winter, Perez (1983) introduced a link between the cyclical theories of technological evolution and the theories of path-dependency, and after Freeman (1987) reintroduced the role of 'institutions' into evolutionary economics (following Veblen 1898, 1919), many new approaches have successfully identified the impact of technological change on the 'structural dimensions' of the economy. In this context the institutional set-up of the economy (broadly defined) guides everyday actions and routines, and it may also be the guidepost for change.

Some economists in the tradition of 'innovation institutionalism' focus on technological trajectories (Dosi 1982, 1988), while others relate economic evolution to the transformations of socio-economic paradigms (Perez 1983, Freeman and Perez 1988). The notion of 'systems of innovation' has also been referred to as special kind of institutions (see 'The systems perspective' below).

A major problem addressed in this book is to derive the instituted processes by which path-dependent structural changes have taken place during the last century of technological innovation governing business opportunities. In such approximation, a technological regime or paradigm provides a framework for how economic processes are instituted into specific evolution paths or trajectories. In this context, the book reveals what the trajectories of technological opportunities (Chapter 3), trajectories of technological innovation growth cycles (Chapter 4), as well as trajectories of corporate technological profiles (Chapters 7 and 8) actually looked like, rather than just assuming their existence.

Special attention is given to irregularities or creative destruction (using Schumpeter's terminology) of paths across technological epochs. (This is explicitly addressed in Chapters 3, 7 and 8, but implicitly addressed in most other chapters.) This may also have implications for how we understand the notion of progress. As this book overall illustrates, although knowledge bases in new systems build upon old ones, their technological scope and limitations, as well as type of business scope, may be totally different.

The 'micro to macro' perspective

Evolutionary economics typically involves an interactive and multi-sectoral focus on change, in order to identify the dynamics of changing structures at a given time. In this context the macro socio-economy is not simply the aggregate of various micro units (as in mainstream reasoning), but is instead regarded as a complex outcome of micro relationships or interactions. To take the notion of structural change as an example, the basic mechanism is one where, as entities grow at different rates, the associated structures change (due to new sorting). This transforms the overall configuration of systems of development, so that the parameters which determine (i.e. select) the directions of development change, facilitating the emergence of new structures, and so forth.

The 'micro to macro' perspective (or the 'micro-based macro') is especially appropriate in a system perspective (see below), as a system is regarded as more than the sum of its parts (or sometimes less, depending on the interaction of micro units). This is related to the holistic approach of the system perspective.

Carlsson, Eliasson, and Taymaz (1997) and Eliasson (1996, 1998) emphasised the 'micro to macro' perspective as the mechanism in which spill-over and evolution takes place. Likewise, Andersen, Metcalfe and Tether

(2000) have suggested how emergence and shaping of innovative activity may be distributed across a wider matrix of structures, interrelationships, constraints, and tradeoffs throughout the economic system.

In the context of this book, developing and changing structures, as well as trajectories of technological change and corporate innovation at the macro level, are not measured directly or treated as a descriptive framework from which reasoning can be drawn. They are instead 'derived' endogenously from the collective interaction and behaviour of micro units.

The system perspective

Especially after Freeman's (1987) first statement of a 'system of innovation', a rich body of ideas of innovation, the nature of the firm, as well as the role of institutions, and methods of science and technology policy have been developed. Although there is no connection with the systems thinking more generally in the social sciences, or with the cross-disciplinary systems movement established by Von Bertalanffy (1968), many evolutionary economists have a vague and not coherent idea of what they mean by a system.

However, no evolutionary economists have really explicitly addressed what a 'system perspective' is, and what difference it makes to the dynamic analysis of systemic change and development. As at starting point, it was emphasised by Andersen, Metcalfe and Tether (2000) that, in the evolutionary perspective, attention is focused upon populations of actors and the dynamic consequences which flow from their behaviours. By taking the population itself as one unit of analysis, we can treat it as part of a wider system of relationships for the generation of innovation. This is the approach taken in Chapters 5 and 6, where the dynamic processes are analysed from a system perspective.

In a system framework the units, level and boundaries are determined by the particular problem of investigation. The boundaries of the dynamic processes have especially been conceptualised in relation to 'innovation systems' (Lundvall 1992, Nelson 1992), 'technological systems' (Carlsson 1987), or 'distributed innovation systems' (Andersen, Metcalfe and Tether 2000), or conceptualised in relation to system-based 'micro to macro' models of the economy (as mentioned above). Hence, the level of analysis is flexible, distributed across all meso, macro and 'micro-based macro' factors, that contribute to the development and transformation processes of the population of investigation.

Systems are constituted by their interdependent and interactive members (or population). In this context, it is here suggested that the principal contribution of a flexible distributed innovation system perspective (as the one suggested by Andersen, Metcalfe and Tether 2000, as well as Coombs and Metcalfe 1998) is to raise the difficult questions about what is it which constitutes a distinct unit of membership, and what the processes are which create the changing structure of this membership, and which explain how systems emerge, grow, decline and disappear. Chapter 5 links the now recognised relationship between the cyclical theories of technological evolution and the theories of path-dependency with the theories of systems, when deriving the changing populations of technologies participating in 'waves' of innovation trajectories. Chapter 6 applies a capability perspective to the dynamics of industries, in which the industry systems' populations are derived from the changing distribution of capabilities of firms within industries, participating in the generation and exploitation of the common knowledge bases. This latter competence bloc view on systemic change is taken from one of the most comprehensive attempts to investigate the systemic aspects of innovation and competition at the microeconomic level, and which is provided by Eliasson (1997) and colleagues.

Relative measures

Schumpeter emphasised that any stage of an economic system (c.f. whether it is in prosperity, recession, crises, depression, or revival) is relative to a certain idea of a 'theoretical normal' level. (This was indeed how he introduced his concept of a 'moving' equilibrium, see Chapter 4 for discussion.)

However, the notion of relativity is taken further in this book, in which the evolution of any entity within a system affects the position of others. This is one of the essential principles when understanding development in a dynamic framework. Hence, the direction of an evolution is a relative concept and should be measured and understood relatively. To take the evolution of technological opportunities in Chapter 3 as an example; technology Y can, for example, retain constant opportunities over time in absolute terms, but given opportunities in other sectors can change over time (the exogenous environment is not constant), the degree and scope of technology Y's relative opportunities changes. Likewise, technology Y's opportunities may change in one direction in absolute terms, but as the exogenous environment is not constant (i.e. other sectors may also change), the degree and scope and hence relative opportunities of technology Y may go in a different direction or not change at all. The same type of reasoning applies to the evolution of corporate competencies in Chapters 6, 7 and 8. Hence, the evolution of the structural dimensions of technological change and corporate competencies, etc., are determined by the relative positions and their distribution.

This view on relativity is incorporated in the theoretical conceptualisation and in the design of empirical measurements throughout the book.

Development, change and emergence

The notion of relativity also implies some existence of inequality or unevenness in a socio-economic system's development patterns. Any such inequality can be at the level of technology's relative scope or advancements, or according to firms' types and degrees of technological competencies, or the whole internal match within a socio-economic system. It is believed that it is the accumulation and catching up between the different unequal entities, which produce system development.

In this book (Chapter 6) it is suggested that dynamic forces within the formation of broader technological systems are created by internal tensions in which economies of scope can be achieved from the accumulation, catching up and combination of different technologies, co-evolving with changing industry structures and boundaries. Such ideas were formerly presented by Hughes (1992), who refers to salients (or reverse salients) in the development phase and to the solving of critical problems. Hughes also points out that some components or sets of components in a technological system may lag behind (or advance in front) of a system as a whole, because they are less (or more) efficient or economical, or in some other way show up in the backwardness (or advanced state) of the other components in the system. An impetus is then given to the system as a whole, in which the most efficient technologies create the development potential for the others. The salient metaphor and the notion of uneven technological system development can also be related to Dahmen's (1989) concept of development blocks (a term which is adopted in Chapter 6) and Dosi's concept of technological paradigms and the evolution of technological trajectories (Dosi 1982, Dosi *et al.* 1988). Rosenberg (1982) writes of bottlenecks in technological development to capture a similar idea to Hughes' 'reverse-salients'. Rosenberg also uses the notion of 'focusing devices' in arguing that technological change historically moves in 'patches'.

Furthermore, if technologies, industrial sectors' corporate technological capabilities, and other variables in the socio-economic system, grow at different rates and with different impact at different times, we will have a change in structures (due to new sorting). This process, in its turn, leads to new combinations and sometimes the emergence of new systems. Evolution, therefore, is not only about catching up and accumulation (i.e. development), but also about new combinations of technologies and industry structures and other variables, and the breaking down of old ones (which relates to Schumpeter's ideas about creative destruction). This is also the idea, for example, behind Kodama's notion of technological fusion and the outgrowth of new industries (Kodama 1986, 1992), and behind the work of Carlsson and Stankiewicz (1991) on the nature, function and composition of technological systems.

Hence, as demonstrated in Chapters 5 and 6, the boundaries, relative scope and overall configuration of a socio-economic system is important for the its' pattern of development, change and emergence, and that systems can act as a barrier as well as stimuli to knowledge accumulation.

Throughout the investigation of processes in this book, focus is mainly on *relative* measures in relation to *development and change* (as that tells something about the dynamic structures at a given time), rather than absolute measures and levels (as that merely tells something about the accumulated past - important as it may be).

Complexity

In discussing interaction processes and interrelatedness, there seems to be a (vague) common understanding among Schumpeterian economists concerning what we mean about the, now frequently used, words complex and complexity, although we have not consulted other sciences to define them or attempted to clarify how the Schumpeterian approach to complexity differs from other approaches. Whereas complexity in Schumpeterian approaches often means that our approach of reasoning has become more complicated by incorporating more links, interactions and structural dimensions, complexity in other approaches, as e.g. in chaos-theory approaches in economics, means high levels of non-linear interactions causing unpredictability. These two approaches are not necessarily coherent. As stressed by Andersen, Metcalfe and Tether (2000, p.29) we have to distinguish between complex and complicated.

This book considers the structural dimensions of technological change and corporate innovation in line with Schumpeterian approaches, while raising questions related to the complexity of development paths, change and emergence.

1.4. THE SCHUMPETERIAN HERITAGE

By exploring some central dynamic processes of change and development, as well as the 'rules' underpinning such, an important step could be to gain insight into an understanding of the *performance* of a specific socio-economic system, including its *sustainability* and its *inherent contradictions*. Such comprehension indeed raises questions with respect to the nature of capitalism, but any such discussion is beyond the scope of this book.

However, this book can be regarded as a step towards an understanding of the evolution of industrialised societies, by addressing the nature and processes of the evolution of technological change and corporate innovation in a historical perspective, as well as the 'rules' underpinning such.

As the nature of such 'problems' follows neither disciplines nor political boundaries, it would be possible to integrate both Schumpeterian, Marxian, as well as Hayekian ideas (all of which are indeed strongly represented within Andersen and Lundvall, forthcoming).

However, this book builds upon the intellectual tradition which probably is the one which dominates evolutionary economics the most today, and which is what Rosenberg has termed 'the Schumpeterian heritage' (Rosenberg 1976, pp.66-68; 1982, pp.5-6).

2. Indicators and Appropriate Use of Patent Data

Abstract

Since the analysis of this book rests mainly upon the information content of patent data, a discussion of the possibilities and problems, as well as the appropriate use of such data, is carried out in this chapter. This is mainly in relation to the variables that patent data serve to proxy. The content and organisation of the US patent database which inform this book is also described.

2.1. INTRODUCTION: BROADENING THE SOCIAL SCIENCES

During the last few decades there has been a growing interest in searching for technology indicators. This is due to the growing recognition of the importance of technology and technical change with respect to economic fluctuations, the competitiveness and growth of firms and countries, as well as historical shifts in world-wide technological leadership and performance. It was recognised that mainstream economic models which treat technology as an exogenous variable could explain only very little about these issues.

While recognising that technology has to be measured and understood just like other factors of production such as labour and capital, there has been an increased interest in using patent data as an indicator in the social sciences. This is due to the fact that, in most mature economies, the exploitation of knowledge and new ideas embodied in product and process innovations have been protected through the use of industrial patents, historically.

Moreover, until now our ability to interpret the evolution of technologies at the firm level has been severely hampered by the lack of suitable indicators of those technologies. This is especially true for historical studies. Patent data have been used extensively in recent studies, but almost invariably these data have been constrained to US patent data for the relatively recent past (i.e. from 1969 or later after which time the US Patent Office has made partially available its raw data on corporate affiliations of patents granted). Other than in a very rudimentary way, the long-term issues have not been clarified. This book will radically change that situation by permitting detailed insight into the corporate structure of patenting over a hundred years period. That is, this book draws upon a US patent database comprising both individual and corporate patents granted in the US by the US Patent and Trademark Office from 1890 and up to 1990, and classified at a very detailed level of disaggregation.

The US patent system is one of the oldest patent systems and had become a distinct bureau under the US government by 1802.

2.1.1. Chapter outline

After first introducing what a patent is, the chapter then moves on to discuss the use of patent data as economic indicators. Special attention will be on justifying the four variables patent data serves as proxy measures for in this book. These are (1) technological innovation, (2) technological opportunity, (3) degree of accumulated technological impact or socio-economic importance, as well as (4) corporate technological capability or competence. The chapter then discusses why this book has chosen a US patent database to provide the historical source for this book. Finally, the organisation of the US patent database on which this book is based will be presented.

2.2. PATENT DATA AS AN ECONOMIC INDICATOR

As in introduction, it is relevant here to introduce what a patent is. The definition of the US Patent and Trademark Office is drawn upon since this book is based upon patents granted in the US:

> "A patent for an invention is the grant of a property right to the inventor, issued by the Patent and Trademark Office. The term of a new patent is 20 years from the date on which the application for the patent was filed in the United States or, in special cases, from the date an earlier related application was filed, subject to the payment of maintenance fees. US patent grants are effective only within the US, US territories, and US possessions.
>
> The right conferred by the patent grant is, in the language of the statute and of the grant itself, "the right to exclude others from making, using, offering for sale, or selling" the invention in the United States or "importing" the invention into the United States. What is granted is not the right to make, use, offer for sale, sell or

> import, but the right to exclude others from making, using, offering for sale, selling or importing the invention."
>
> *US Patent and Trademark Office*
> *http://www.uspto.gov/web/offices/pac/doc/general/whatis.htm*
> *May 2000*

In order for an invention to be patentable it must be 'new' as defined in the patent law, which asserts that an invention cannot be patented if it does not satisfy certain novelty conditions[1]. Thus, as a patent has to reflect a novelty (i.e. a movement of the technological frontier), it has often been used as an economic indicator when measuring the rate and direction of technological change and corporate innovation. However, the criteria for patentability is not strait forward:

> " .... the troublesome question of what ideas should be granted patent protection must be faced. In one extreme, there is nothing new under the sun. In the other extreme, every different combination of ideas or every different application of an idea constitutes a new idea. In specifying the criteria of patentability, the designers of any patent system must select a position somewhere on the spectrum marked by these extremes."
>
> *S. N. S. Cheung (1986, p.6)*

'Novelty', in comparison to what is a recognisable combination of the existing, has become even harder to identify with new technology, such as information technology, making it possible to merge and change existing scientific and engineering compositions. The criteria for 'IPRability' differs across countries, and this hampers a direct comparison of the propensity to patent across different national systems. Archibugi (1992), for example, has revealed how it appears there are more 'new under the suns' in Japan than in the US (to put it in Cheung's phrase) as more is patentable in Japan.

Furthermore, 'law of nature' such as gravity, or the discovery of the already existing (such as new species), are not patentable. However, this is not straight forward either, as we can see from the recent discussions concerning the extent to which we should be able to patent genetic codes.

This chapter does not aim to argue the relative merits of various different technological innovation indicators (as e.g. data on R&D expenditures, bibliometric data and data on productivity changes) in comparison to the use of

---

[1] See 'Novelty and other conditions for obtaining a patent' by the US Patent and Trademark Office: http://www.uspto.gov/web/offices/pac/doc/general/novelty.htm (May 2000).

patent data, except to mention that patent data are probably the most long running detailed historical record of technological activities, and are therefore very suitable to use as an indicator in historical studies and approaches such as the one applied in this book. For example, many, especially small, firms do not have formal R&D divisions and this is especially true in a historical perspective, so expenditures on R&D are not formally registered historically. Also, as R&D is an input variable which is related to downstream innovation with a great deal of uncertainty, no strict relationship between R&D expenditures and innovation activity has been observed. However, Mansfield (1986) has shown that firms apply for a patent for about 66% to 87% of their patentable inventions, while Scherer (1959), Sanders (1964), Napolitano and Sirilli (1990) have shown that 40% to 60% of total patent applications actually progress to innovations. Of course, as pointed out by Archibugi (1992, p.359), although empirical evidence seems to encourage rather then discourage the use of patent statistics, one has to be aware that universities overall have another attitude towards their research output, and that they are more likely to generate papers in scientific journals rather than patent applications. Hence, one could alternatively choose to use bibliometric data in searching an indicator of technological innovation, but such data tend to be more related to basic science rather than invention and innovations. Also in a study which aims to measure firms' relative technological specialisation, bibliometric data is not appropriate, since firms tend to patent rather than publish in scientific journals. Furthermore, a patent has to reflect a novelty (i.e. a movement of the technological frontier, as mentioned above) and is therefore an appropriate indicator when measuring the rate and directions of technological change (as mentioned above), whereas such novelty restrictions are not imposed on scientific publications to the same extent. Finally, neither R&D data and data on productivity change can measure the disaggregated pattern of technological change or the technological specialisation of firms by comparison with the patent output variable, which is classified at a very detailed level. Hence, patent statistics offer a potentially very rich source of empirical evidence on questions related to the structure of technology, and how it changes over time across countries, industries and firms.

Moreover, in recent times the effectiveness and social influence of patent data recording has increased. With improvements in information technology, national policy makers, international agencies, industrialists and researchers can now for their own individual purposes on-line systematically retrieve the information from national patent offices. For example, the US Patent and Trademark Office have almost all their publications and information online as well as in printed form. These publications contain a vast amount of general information in a non-technical language intended especially for inventors, prospective applicants for patents, and students (www@uspto.gov). The information contained in the patent documents concern both the patentees and the patented inventions themselves at a very detailed level (including information concerning search on patent groups, classes and subclasses), as well as other material. Therefore, use of others' patents in own research (including cross-licensing) as well as applying for a patent has become more accessible. Hence, the increased use of patent records has increased the usefulness of patent statistics. Also, as mentioned by Archibugi (1992, p.358) a number of companies today provide on-line information on the patent literature, and several newsletters, such as *World Patent Information*, diffuse information on new methods of patent search. Likewise, Pavitt (1984a) also emphasises how advances in information technology have increased actual and potential uses of patent statistics.

Hence, it is believed that social science, especially in a historical perspective, will be more comprehensive by supplementing (not substituting) technology studies with the use of patent statistics. Therefore, for historical reasons, and due to the increased influence of detailed patent recording, patent data have been chosen as the source of reference when attempting to contribute to knowledge concerning processes of technological change and corporate innovation.

Although the use of patent data in social research has gained attention, it is not a new idea to use patent data as an economic variable. Patent statistics have now proven achievements of promise in analysing: (i) international patterns of innovative activities and their effects on trade and production, (ii) patterns of innovative activities amongst firms, and their effects upon their technological strength or competence as well as performance and industrial structure, (iii) rates and directions of innovative activities in different technical fields and industrial sectors, and (iv) links between science and technology. Many of those studies will be referred to in this book.

However, only very few historical studies have exploited the panel data characteristics that patent statistics can provide when classified by type of technological activity, corporate ownership, industry of firm, as well as over time. Most longer term historical studies of trends in patenting (such as those of Beggs, 1984; Kleinknecht, 1987; or Brouwer, 1991) have relied principally on aggregate statistics, with very little cross-sectional subdivision, which open up for an evolutionary perspective. Hence, while investigating technological change and the evolution of corporate innovation, this book will explore some of the panel data characteristics patent data can provide when sorted at a very detailed disaggregated level.

It is of course believed that patent data can be used and misused or abused as any other data source used in statistical studies, and that patent data is not appropriate for all kinds of research. Possibilities and problems of patent statistics have also been discussed in many patent studies. For an overview, see the surveys by Schmookler (1950, 1953, 1966), Reekie (1973), Pavitt (1984a,

1988), Narin *et al.* (1987), Griliches (1990) and Archibugi (1992). This chapter does not aim to contribute to the overall survey literature concerning use of patent data. Instead, possibilities and problems including appropriate use of patent data will be discussed *only* in relation to the issues addressed in this book.

This book is related to the above indicated areas in which use of patent data has proven promise in social science. The first part of the book will use patent statistics and panel data analysis to contribute to the discussion concerning determining the processes of technological change. The analysis will be related to the evolution of technological opportunities, innovation growth cycles, and secular movements in development patterns. Patent statistics in the last part of the book will serve as an indicator when investigating the processes of corporate technological competence development on an industry and firm basis. The analysis will be related to industry knowledge bases, corporate technological profiles as well as to innovation diffusion.

In this book patent data serves as an indicator for four interrelated variables:

1. 'Technological innovation'.
2. 'The extent of technological opportunity' which is reflected in the rate of technological innovation.
3. 'Accumulated technological impact or importance', which is acquired from innovation over time, and hence reflected in the stock of such.
4. 'Corporate technological capability or competence', which is reflected in the degree and type of corporate technological specialisation.

As patent data is only a direct measure of invention, there are equally *four potential difficulties* with this approach in which patent data serve as proxy measures:

Concerning 1. Patent data as a proxy for technological innovation

It is argued here that patent data under certain circumstances can serve as a proxy for technological innovation, and that the evolution of the structure of patenting in that way can be regarded as reflecting the underlying pattern of technological change. For the purpose of justifying the relevance of this assumption, it is here suggested that cumulative calculations (stocks) are especially appropriate when patents are used as a proxy measure for technological innovation, as that measure captures all the features of innovated new technology. In this context it is argued that accumulated inventive activity of course does not take place unless applied, so that accumulated patent stocks reflect the process of application and diffusion of innovated new technology (i.e. the interaction-process between scientists, inventors, engineers, innovators and entrepreneurs, learning and market diffusion). Hence, by using accumulated patent stocks this book will take an appropriately broad view of technology, where it includes the interaction between (i) the universal element of technology relating to information or codified knowledge (for example patents) which is potentially tradable and potentially transferable, and (ii) the tacit element of technology which is context specific and tied to local technological competence or capability, as well as (iii) that of market diffusion. In that way (i) may serve as a proxy for (ii) and (iii), as well as being a direct measure of itself.

Hence, cumulative stocks of patents have been calculated for technological sectors, technological groups, broader technological groups and total. Stocks were calculated using the perpetual inventory method as in vintage capital models, with an allowance for a depreciation of the separate contribution of each new item of technological knowledge over a thirty-year period - the normal assumption for the average lifetime of capital, given that new technological knowledge is partly embodied in new equipment or devices, as well as the conventional method used in patent statistics, as introduced by Cantwell and his associates[2] who pioneered the use of accumulated stocks of long time series within patent statistics.[3] Thus the stock in 1919 represents a weighted accumulation of patenting between 1890 and 1919, a thirty-year period with weights rising on a linear scale from 1/30 in 1890 to unity (30/30) in 1919, using 'straight line depreciation'.

Although the assumption of a thirty-year life is admittedly arbitrary, it must be emphasised that this is a proxy measure of the life of the underlying *technological knowledge* and the tangible devices with which it is associated. Hence, it is *not* intended to capture lifetime of the patent (which is shorter) or the life of the economic value of the patent.

However, the results would be largely unaffected if a shorter lifetime was assumed. Although patent stocks (used for analysis throughout the book) would then fluctuate more as the smoothing process associated with accumulation would be less pronounced, and so the absolute values of the growth of stocks and the inter-corporate dispersion of activity would be greater, the identification of the periods in which stocks grow relatively faster or slower would be largely unaffected. Also, the identification of innovations cycles (in Chapter 4), as

---

[2] Accumulated stocks of long time series within patent statistics was e.g. applied by Andersen 1998, Cantwell and Andersen 1996, and Fai and von Tunzelman forthcoming concerning technology and industry dynamics issues. Cantwell and Fai 1999 applied it to firm dynamics issues.

[3] Previous work on long time series with patents was of course done by Kuznets (1930a) who accumulated stocks only over 5-year periods and investigated the change in those, and by Schmookler (1966) who worked with flows.

opposed to random fluctuations, depends upon the cut-off points between what is recognised as a cycle and what is taken as a random fluctuation on the cycle, and *not* by the rate of depreciation.

The use of patent stocks is also consistent with the theoretical notion of technological accumulation (which again is analogous to capital accumulation), and which follows from the view that technological change is a cumulative, incremental and path-dependent process (Nelson and Winter 1982, Rosenberg 1976, 1982, Dosi *et al.* 1988).

Analysing stocks also helps to reduce statistical problems that might otherwise be created by small numbers of patents and by year-to-year fluctuations in patenting, problems which are more serious at more detailed levels of disaggregation, and which cause only random results. In order to avoid problems related to these issues, some restrictions have been imposed in the design of empirical categories and methods when selecting technological sectors and corporate patenting for analysis in the different chapters.

Concerning 2. Patent data as a proxy for technological opportunity

The second potential difficulty is that patent data serves as a means of identifying the extent of technological opportunity in a sector, derived from the rate of growth of total stock of technological capability. It is here assumed that a fast rate of growth of patenting (i.e. a high rate of growth in patent stock) in one technological sector or class of activity represents an area of strong technological opportunity in the period in question. This is now a standard assumption within patent statistics in general.

However, for the purpose of justifying the relevance of this assumption, quantitative evidence of patent statistics, as to which technological fields have enjoyed the greatest new technological opportunities in the twentieth century, has been married up, compared, and found to be consistent with the history of technology literature which provides an alternative qualitative assessment of trends in the technological evolution over the same periods across different technological fields, as well as other case studies of technologies chosen for their particularly important contributions to development (Andersen 1997). Some of such comparisons are also presented in Chapter 6 of this book. Historical patent data makes such cross-checking easier (which then becomes *ex post*) than what is possible in contemporary studies of patent classes which are currently the fastest growing, in which the assumption that these classes depict the fields of the greatest new opportunities can only be established by an evaluation based on the judgement of experts in these fields.

Concerning 3. Patent data as a proxy for accumulated technological impact or importance

It is here argued that patent data can serve as a proxy for accumulated technological impact or socio-economic importance. A potential problem when using patent data for such measure is that the value of each patent might be very heterogeneous. Some patents reflect an area of significant impact while other patents are never used. However, if for example a patent in a certain area is never used or of no importance, the accumulated patent stock within that technological field will not grow and therefore not reflect an area of great technological opportunity. Hence, technological impact is reflected in a succession of innovations over time, so that the size of accumulated patent stocks is positively associated with areas of great technological impact historically.

Also, studies of Cantwell (1991) and Cantwell (forthcoming) have found that the most serious drawbacks that are involved in the (inappropriate) use of patent statistics - most notably, stochastic fluctuations in variations in the importance of individual patents (and in the propensity to patent across sectors and over time) are substantially ameliorated over large numbers of patents. It is believed here that by working with a large number, the relative importance within the whole population of the patent data tends to follow a normal distribution.

Finally, it is also argued here that the overall effect of this heterogeneity in the results is further minimised by especially working with *rates of change in patent stocks* rather than absolute flows which have been commonly used elsewhere.

Concerning 4. Patent data as a proxy for corporate technological capability or competence

At last, it is supposed that the pattern of corporate technological specialisation at a very detailed level serves as an indicator of corporate technological competence or strength. Therefore, corporate patents will be used as a proxy for the underlying pattern of corporate technological specialisation which in turn reflects the distribution of technological competence across firms.

Patents are obtained by firms for several reasons. They may e.g. be obtained by firms (and individuals) for strategic reasons, e.g. in order to block competitors, and they may be obtained in order to introduce innovations through cross-licensing (like exchanging cooking recipes in order to increase the range of future opportunities and stimulate faster learning). There may also be other reasons (or rationales) of obtaining a patent (see section 2.3. and Figure 2.1). However, no matter what the primary reason is for a particular firm to obtain patents, they still represent a technological capability.

In order to minimise the well known problem in patent statistics that, there is a difference amongst firms across different industries in the propensity to patent the output of innovative activities, the degree of firms' technological competence in selected technological fields is measured and compared only in relation to other firms within their industry (except for in Chapter 6, where the purpose of investigation is to identify the extent to which firms have increased their patenting across industry boundaries historically). Yet, overall regularities of changes in the pattern of companies' technological profiles, can of course still be compared across industries (cf. Chapters 7 and 8).

Moreover, as some firms are larger than others, smaller firms naturally more easily may become highly specialised in a specific technological field in absolute terms in comparison to larger ones; and it is also well known that larger firms are more technologically diversified than small firms (Chandler, 1990). Hence, in order to adjust for the bias this may create in the results, the types of technological competencies will be defined in such as way that what matters is neither the firms' overall technological size nor their overall degree of specialisation, but rather the actual composition or profile of their specialisation - that is, whether their resources are especially geared towards development in certain technologies (e.g. fast growing or core) relative to other firms within their industry. Hence, smaller firms may have a certain type of technological competence if they are particularly oriented towards development in certain technological fields, although the absolute level of their activity in these fields may be less than that of giant companies.

Finally, the overall results concerning firms' relative technological competence may to some extent be affected by firms with very little patenting (so the pattern of patenting within such firms is influenced to a significant degree by chance circumstances, as opposed to the real underlying distribution of technological activity). Therefore, in the design of empirical categories and methods, and in the selection of firms for specific analysis, attention is paid to the difficulties that might be created by firms with little patenting.

2.3. PATENTED IN THE US

When examining technological change and the evolution of corporate innovation in a global context using patent data, the next potential difficulty is that there is a difference amongst countries in the economic costs and benefits associated with patent protection. This reflects the costs, time and rigour of the patent examination and enforcement procedures, together with the expected size of the market. Also, there is a difference in governance of patent systems across countries (Andersen and Howells 2000), as will also be demonstrated in what follows. It has been argued that patents granted in the US provide the most useful basis for international comparisons, given the common screening procedures imposed by the US Patent and Trademark Office (Soete 1987; Pavitt 1988).

Moreover, as the US historically has been a progressive economy with the world's largest single market during the last century of technological development and expansion, and a country that explicitly encouraged and welcomed new ideas and innovation, successful inventions (home or abroad) were likely to be patented there[4].

Furthermore, it has been argued by Rosenberg that technological change has been a critical variable in accounting for the spectacular long-term growth of the American economy (Rosenberg 1982: Chapter 3). Hughes (1989) refers to a 'second discovery of America' which began during the opening decades of this century:

> "The first discovery had been that of the virgin land, nature's nation; the second discovery was of technology's nation, America as artefact. ... for more than a century the nation had been the world's most active construction site. ... European intellectuals [scientists and philosophers], architects [design, engineering and construction], and artists [help to articulate the modern culture that modern technology inaugurate] making the second discovery believed that the United States was leading the world into a uniquely modern era."
>
> *T. P. Hughes (1989, p.295).*
>
> *Note: Comments in square brackets indicate own interpretation*

However, as the intellectual property right (IPR) regime (including the patent system) participate in shaping innovation strategies and innovation practices in which individuals and firms appropriate, it is important not just to assume the existence and nature of any such regime, but to ask the question why we have it in the first place. Andersen and Howells (2000) grouped some different rationales (associated with IPR 'system believers').

---

[4] 'For over 200 years, the basic role of the Patent and Trademark Office (PTO) has remained the same to promote the progress of science and the useful arts by securing for limited times to authors and inventors the exclusive right to their respective writings and discoveries (Articles 1, Section 8 of the United States Constitution). ... Through the issuance of patents, we encourage technological and advancement by providing incentives to invent, invest in, and disclose new technology worldwide. ... By disseminating both patent and trademark information, we promote an understanding of intellectual property protection and facilitate the developments and sharing of new technologies world wide.' (www@uspto.gov. Last modified February 1996.)

Figure 2.1. The rationales to IPR systems

Moral rationales	Human rights:	The law should provide remedies against those who appropriate ideas of others, and a person who has devoted time and effort to create something has a right to claim the thing as his own, and also has a right to obtain some reward to all his work.
	Business ethics:	IPRs function as a safeguard for consumers against confusion of products and quality as well as deception in the marketplace (this indeed applies mostly to trademarks).
Economic investment rationale	Incentives to creativity:	IPRs provide the prospect of reward which in turn encourages creative and technological advancement by increased incentives to invent, invest in, and innovate new ideas. (Say 1834; Mill 1862; Clark 1907).
	Market creation:	Efficient IPR protection allow profit oriented firms to enter (or develop) an industry or market.
	Increased competition:	IPR helps to cover the fixed costs of inventing and producing a new product as well as protecting against new marked entry. This may stimulate a creative dynamic environment as well as strengthen and broaden continuous innovators.
Economic rationale of organising science, technology and creativity	Order:	It has been argued that 'The prizes of industrial and commercial leadership will fall to the nation which organises its scientific forces most effectively' (Elihu Root in Noble 1979, p. 110). (This view was mainly raised with respect to establishing scientific laboratories, but it is here also suggested to apply to an adequate science and technology system organised at the nation level.)
	Increased information and spill-over:	IPRs facilitates the developments and sharing of new technologies and creative efforts world wide. Patents and copyrights, when filed, provide immediate information to rivals who can incorporate such into their own knowledge bases even though they cannot make direct commercial use of it. This might create a more coherent technological and industrial development, faster spill-over in knowledge and creative efforts and technological progress which strengthens the national or global economy.
	Increased information and better advice:	An intellectual property system also offers information concerning structural changes in technological development as well as technological capabilities of industry and sectors, allowing governments to be more effectively advised on science and technology policy matters.
	Uniformity:	A national system brings in national uniformity (as opposed to regional differences in IPR legislation). This makes it possible to (or seeks to) promote cross-country trade in IPRs and international integration of science, technology and creative efforts, stimulating prosperity world-wide.

Source: Revised version of Andersen and Howells (2000).

From Figure 2.1 we can see how the introduction of IPR regimes have been associated as much with the underlying moral, ethical, social and philosophical issues of intellectual property as its direct economic ones. These institutional structures may in turn reflect the different underlying rationales that nation-states have applied to the introduction and development of IPRs. See Figure 2.2, as a matter of comparison of the US system with other systems.

Figure 2.2. Government departments under which 'patents' are administered

US	UK	Germany	France	Japan
Department of Commerce	Department of Trade and Industry	Federal Ministry of Justice	Ministry of Culture and Francophone Affairs	Ministry of International Trade and Industry

Firstly, all countries with an effectively developed IPR regime must be regarded as IPR system believers concerning believing in some reward of organising science, technology and creativity. More specifically, it could be argued that patents systems in the US, Japan as well as UK illustrate an overall 'economic rationale' being allocated under the Department of Commerce or Department of Trade and Industry. By contrast, in Germany the patent system is administered by Federal Ministry of Justice which suggests an emphasis on a 'moral and ethical' rationale based on 'rights'. Finally, concerning France, having the patent system administered by the Department of Culture and Francophone Affairs, may reflect a strong historical moral and ethical rationale basis of protection of intellectual creativity.

As the IPR system of different countries is strongly historically rooted and embedded in often very different institutional structures, it is also most appropriate to base the analysis in this book on patenting within only one country. In this context, US is the economy with the most developed property right system (including its enforcement) historically (Andersen and Howells 2000). Besides, the macroeconomic efficiency of governance of any patent (or IPR) regime, as well as its shaping element for innovations at the micro level, also reflects the usefulness of using patents as indicators when exploring innovation dynamics.

Consequently, for several reasons it is argued that data derived from patents granted in the US provide the most useful indicator for identifying technological change and the evolution of corporate innovation historically.

2.4. THE US PATENT DATABASE ON WHICH THIS BOOK IS BASED

Accordingly, this book is based upon a patent database comprising both individual and corporate patents granted in the US from 1890 and up to 1990. It has been compiled by Professor John Cantwell at the University of Reading, with the assistance of the US Patent and Trademark Office.

Each patent is classified by the year in which it was granted and by the type of technological activity with which it is most associated grouped at different levels of aggregation. The country of the inventor (individual or corporation) is also listed, as well as the country in which the patent is granted (which is always US in this case).

Information was collected manually from the *US Index of Patents* and the *US Patent Office Gazette* on all patents that were assigned to a selection of large US-owned and European-owned firms between 1890 and 1969. From 1969 onwards equivalent information has been computerised by the US Patent and Trademark Office and incorporated into the database, and the US Patent and Trademark Office also provided a technological classification of all patents granted from 1890 onwards at a detailed level of disaggregation.

Various broad categories of technological activity can be derived by allocating classes or sub-classes to common groups of activity. For this purpose patents have been allocated to one of 399 technological sectors, which in turn belong to one of 56 technological groups. In this context, patent classes (or occasional a subdivision of a patent class) that fall within one of the 399 technological sectors have been assigned to one type of technological group, while other patent classes (or a subdivision of a class) have been assigned to another type of technological group. For example, some of the sub-classes that fall within US patent class 62, 'refrigeration', comprise a sector that has been assigned to the technological group 'chemical processes', while the remaining sub-classes within 'refrigeration' constitute a different sector which has been allocated to 'general electrical equipment'. The 56 categories have been amalgamated into five broader groups of activity consisting of Chemical, Electrical/electronic, Mechanical and Transport technologies, plus a residual consisting of other mainly non-industrial technologies. Of the 399 technological classes (or occasionally a sub-division of a class), 57 are in the Chemicals group, 64 fall under the Electrical/electronics category, 221 are allocated to the Mechanical group, 21 are allocated to the Transport group and the remaining 31 are allocated to the residual group of non-industrial technologies.

The 56 technological groups, sorted by broad technological category, are presented in Table 2.1.

Table 2.1. Identification of the 56 technological groups

Broad technological group	Code	Technological type
	2	Distillation processes
Chemicals	3	Inorganic chemicals
	4	Agricultural chemicals
	5	Chemical processes
	6	Photographic chemistry
	7	Cleaning agents and other compositions
	8	Disinfecting and preserving
	9	Synthetic resins and fibres
	10	Bleaching and dyeing
	11	Other organic compounds
	12	Pharmaceuticals and biotechnology
	51	Coal and petroleum products
	55	Explosive compositions and charges
	30	Mechanical calculators and typewriters
Electrical/electronics	33	Telecommunications
	34	Other electrical communication systems
	35	Special radio systems
	36	Image and sound equipment
	37	Illumination devices
	38	Electrical devices and systems
	39	Other general electrical equipment
	40	Semiconductors
	41	Office equipment and data processing systems
	52	Photographic equipment
	1	Food and tobacco products
Mechanical	13	Metallurgical processes
	14	Miscellaneous metal products
	15	Food, drink and tobacco equipment
	16	Chemical and allied equipment
	17	Metal working equipment
	18	Paper making apparatus
	19	Building material processing equipment
	20	Assembly and material handling equipment
	21	Agricultural equipment
	22	Other construction and excavating equipment
	23	Mining equipment
	24	Electrical lamp manufacturing
	25	Textile and clothing machinery
	26	Printing and publishing machinery
	27	Woodworking tools and machinery
	28	Other specialised machinery
	29	Other general industrial equipment
	31	Power plants
	50	Non-metallic mineral products
	53	Other instruments and controls

Table 2.1. Continued

Broad technological group	Code	Technological type
	42	Internal combustion engines
Transport	43	Motor vehicles
	44	Aircraft
	45	Ships and marine propulsion
	46	Railways and railway equipment
	47	Other transport equipment
	49	Rubber and plastic products
	32	Nuclear reactors
Non-industrial	48	Textiles, clothing and leather
	54	Wood products
	56	Other manufacturing and non-industrial

Note: 399 technological sectors are spread across 56 technological groups

As the system of patent classes used by the US Patent and Trademark Office changes, fortunately the US Patent and Trademark Office reclassifies all earlier patents accordingly, so the classification is historically consistent. This book is based on the US Patent and Trademark Office's 1990 classification scheme. Furthermore, although the US Patent and Trademark Office have assigned most patents to more than one technological field or class, the Office identifies the most important or primary class of every patent, and in this study the primary classification was used in all cases.

However, a potential difficulty related to the classification scheme is that, like all classification schemes, the patent class system is to some extent arbitrary. There are occasions in which the patent class system is either over-elaborate (such as in certain classes of drugs and of synthetic resins, in which the divisions between classes may represent administrative convenience as much as genuine technical distinctions) or insufficiently differentiated (as in the case of agriculture chemicals, in which it is difficult to separate fertilisers from pesticides). Thus, caution may be needed in interpreting the profile of corporate activity within those classes that are not quite so precisely delineated than are others as distinct technical fields. However, as a general picture, the classification scheme provides a reasonable differentiation of distinct technical fields.

The firms selected for the historical patent search in this book were identified in one of three ways. The first group consisted of those firms which have accounted for the highest levels of US patenting after 1969; the second group comprised other US, German or British firms which were historically among the largest 200 industrial corporations in each of these countries (derived from lists in Chandler 1990); and the third group was made up of other companies which featured prominently in the US patent records of earlier years, including all US firms with at least 30 patents per year and all European firms granted at least five patents per annum (a method that proved most significant for a number of French firms that had not been identified from other sources). In each case, patents were counted as belonging to a common corporate group or firm where they were assigned to affiliates of a parent company. In total, the US patenting of 857 companies or affiliates were traced historically and together they comprise 284 corporate groups or firms. Births, deaths, mergers and acquisitions as well as the occasional movement of firms between industries (sometimes associated with historical changes in ownership) have been taken into account. The analysis in this book is at the 284 corporate group level and they are classified into 20 industrial sectors (presented in Table 2.2).

Each corporate group was allocated to an industry on the basis of its primary field of production. These industries have been combined into four major industrial groups on the basis of the types of technology that have been most characteristically developed by the firms in the industry historically. Of the 284 corporate groups in all, 82 are in the Chemicals group, 45 firms fall under the Electrical/electronics category, 107 are allocated to the Mechanical group, and the remaining 50 to Transport.

Table 2.2. The organisation of industrial sectors

Broad Industrial Group	Code	Industrial sector
	4	Chemicals n.e.s.*
Chemical	5	Pharmaceuticals
	13	Textiles and clothing
	18	Coal and petroleum products
	8	Electrical equipment n.e.s.
Electrical/ electronics	9	Office equipment
	1	Food
Mechanical	2	Drink
	3	Tobacco
	6	Metals
	7	Mechanical engineering n.e.s.
	14	Paper products n.e.s.
	15	Printing and publishing
	17	Non-metallic mineral products
	19	Professional and scientific instruments
	20	Other Manufacturing
	10	Motor vehicles
Transport	11	Aircraft
	12	Other transport equipment
	16	Rubber and plastic products / tires

Note: 284 corporate groups or firms are spread across 20 industries.
** n.e.s. : not elsewhere specified.*

The sectoral classification of patents, according to the type of technological activity with which each patent is associated, must be distinguished from the industry of the firms to which patents may be assigned, both of which have been recorded separately. Most large firms have engaged in at least some development in most of the broad technological groups of activity, irrespective of the industry in which they operate (see Chapters 6 and 7). Thus, in this use the term a technological sector, which refers to a technological field, is quite distinct from the use of the term industry, which refers to groups of firms with similar products or serving similar markets.

Finally, when investigating the processes of corporate innovation, no distinction is made between product and process innovation. This is mainly because a process innovation for some firms are often a product innovation for others, and *vice versa*, due to most major firms today are horizontally and vertically integrated. Furthermore, as this book is based upon a patent classification scheme, and in the design of empirical measurements, we do not consider the form of each patent *per se*. That is, we only measure technological and innovative activity in certain technological areas rather than specific product or process innovations.

3. Structural Changes in Trajectories of Technological Opportunities[1]

Abstract

This chapter examines the evolution of technological opportunities historically. By analysing the complexities behind changing technological opportunities, technological 'regimes' governing the evolutionary process are traced empirically. The chapter also explores how the composition of technological opportunities has evolved across historical epochs. Evidence shows how the path-dependent evolution of technological opportunities is characterised by 'creative, incremental, accumulation', as it is revealed how new technological systems with new opportunities build upon (complement and extend) the knowledge embodied in old ones, rather than substitute it. Evidence also reveals how the evolution of technological opportunities has become increasingly interrelated, wider-ranging and complex, in which trajectories previously following isolated channels of development are brought together. Finally, evidence shows how typical technological trajectories of broad technological groups explain technological evolution better than the conventional aggregate measures of such broad technological fields.

3.1. INTRODUCTION: THINKING IN PARADIGMS

It has been argued by institutional economists that changing technological opportunities along trajectories governed by technological paradigms or regimes is perhaps the most central regulating variable in society (Dosi 1982, 1988; Perez 1983; Freeman, Clark and Soete 1982; Freeman and Perez 1988). However, any attempts to identify quantitatively and statistically the underlying technological trajectories and patterns of technological development, as well as the establishment of technological paradigms, have lagged behind more qualitative accounts.

[1] This chapter is a revised version of Andersen (1998).

Hence, although it is now commonly agreed among institutional evolutionary economists how important it is to understand the impact of technology on the dynamics of the economy, and that theory must pay special attention to the origin and history of new technologies as technological development is accumulative, incremental and path dependent, it is mainly only among historians that a growing interest has been aroused in studying the changing opportunities of science and technology. Yet, due to the colossal transformation technology makes to our lives, it is important not just to recognise the enormous range of new tools and techniques, products and processes, and new sciences and disciplines developed, but to grasp the underlying trends.

This chapter aims to identify quantitatively, and statistically measure, what has been termed 'changes in technological opportunities' during the last century of technological evolution (1890-1990), in order to subsequently trace empirically the evolution of trajectories of technological opportunities governed by technological paradigms or regimes.

3.1.1. Chapter outline

The selection and organisation of the patent data on which this chapter is based will first be presented. After that, the theoretical framework in relation to the main objective of this chapter will be introduced. The areas of greatest technological opportunities within different epochs of technological development will then be calculated, and subsequently used to extract the complexities behind the changing technological opportunities within and between broad technological groups (Chemicals, Electrical/electronics, Mechanical, Transport and Non-industrial) historically. The purpose of this is first to sketch technological epochs of structural changes in patenting patterns governed by the evolution of technological paradigms.

Then, there will be an investigation of the nature of the changing composition of technological opportunities; whether it changes gradually or is relatively stable, or whether it tends to fluctuate across historical epochs of development. The extent to which changes in technological development causes divergence or convergence in technological opportunities historically will also be the subject for investigation. This will enable an examination of the extent to which paradigms governing new technological epochs 'creatively' destroy old ones or complement and extend them.

Finally, different possible trajectories of technological opportunities will be set out, and each broad technological group's relative contribution to specific paths will be calculated. This will enable us to identify typical (or revealed) technological trajectories of great importance for each broad technological group (i.e. Chemical, Electrical/electronic, Mechanical and Transport).

3.1.2. Data selection

Changing technological opportunities will be calculated from cumulative stocks of patents 1920-1990, as identified in Chapter 2. The analysis is carried out at the 399 technological sector level and in aggregate for each of the five broad technological groups: Chemicals, Electrical/electronics, Mechanical, Transport and Non-industrial. In the design of empirical categories and methods, further restrictions have been imposed when selecting technological sectors for specific analysis. They are presented, where appropriately, in the text.

3.2. THEORETICAL FRAMEWORK: THE INSTITUTED NATURE OF TECHNOLOGICAL DEVELOPMENT

Although 'institutional thinking' has revitalised economics and opened many new approaches to the subject, a common idea concerning the concept of 'institutions' in the institutional tradition within economic theory is that the behaviour of the evolution of societies and technology are characterised by regularities which are specific to time, place, economic sector, and technological field. In this context the institutional set-up of the economy (broadly defined) guides (or sets of 'rules' of) everyday actions and routines, and it may also be the guidepost for change. Thus, the behaviour of all micro units within the economy are 'instituted' (Veblen 1898, 1919; Nelson and Winter 1982; Hodgson 1988; Johnson 1992).

The major source of inspiration in this chapter is provided by the Schumpeterian evolutionary economic approach to institutional economics. In Schumpeter's earlier work he focused on the 'individual' entrepreneur as the most important agent bringing innovations to the economic system (Schumpeter 1934), but later he revised his theoretical scheme by giving a critical role to 'collective' work in the R&D laboratories (Schumpeter 1942). However institutional economists have taken this further and argue that the economic structure and institutional set-up form the framework for a process of interactive learning which sometimes results in innovations. Some institutional economists in the tradition of evolutionary economics refer to 'systems of innovation' in which they point to the existence of 'collective entrepreneurship' (Lundvall 1992, pp.9-10), while other institutional economists focus on technological trajectories and paradigms as special kind of institutions (Dosi 1982, 1988; Perez 1983; Freeman, Clark and Soete 1982; Freeman and Perez 1988).

Dosi (1982, 1988), Perez (1983), Freeman, Clark and Soete (1982), Freeman and Perez (1988) relate economic evolution to the transformations of technological paradigms. Whereas Dosi (1982) introduced a parallel between modern philosophy of science (which suggests the existence of scientific paradigms derived mainly from Kuhn (1962) and the evolution of technological paradigms in which he groups technological discoveries, Perez (1983), in her notion of 'techno-economic paradigms', introduced a link between cyclical theories of technological evolution and the theories of path-dependency and structural and institutional changes (for an overview see Freeman 1994, pp.482-488). An important feature of the technological (or techno-economic) paradigms is that it is not that any direction of technological development can happen, as the technological evolution is instituted into certain paths of development. That is, technological change and progress take place along economic and technological tradeoffs defined by the paradigm. Evolutionary institutional economists point to the existence of technological trajectories. Some trajectories are more likely to be followed than others, as defined by the existing paradigm. Also here there has to be distinguished between those trajectories which are specific to a particular technology, product or industry and those which are of general importance (Nelson and Winter 1977).

Freeman and Perez (1988) differ from Dosi (1988) in the sense that they refer to the Schumpeterian type of *meta*-paradigm of a dominant technological (or techno-economic) regime which rules for several decades, whereas Dosi refers to a type of (micro) paradigm or trajectory which is specific to a particular technology. In Freeman and Perez's framework a radical change in the whole economy is related to a *generalised* shift in the technological (or techno-economic) paradigm, revealing an overall shift in structures at the micro level as well as the macro level throughout the economic system. Accordingly, Freeman and Perez refer to generalised structural changes as the hard core of long-term theorising.

The emphasis in this chapter will only be on inventions and innovations which have diffused and led to generalised technological and economic changes, so this chapter operates within the framework of *meta*-paradigms. It is argued here that *ex post* the successful areas of new technological opportunities - or perhaps more correctly, those which hold up - reflect not just the technological opportunities which have governed and been governed by the paradigms, but also the areas in which society possessed socio-economic competence, as without that the paradigm would not have been unleashed. Hence, although the focus is narrowed to the technological features of the evolution of paradigms, the characteristics of technological epochs - divided by structural changes in the pattern of evolution of technological opportunities - can be regarded as a reflection of the overall features of the paradigm.

In the context of this chapter, the way in which overall technological changes take place (i.e. the pattern of evolution of the technological opportunities which sets the directions of development) are referred to as technological epochs; and they are, as mentioned above, separated by structural changes as well as governed by the evolution of the overall technological paradigms. The specific development paths of opportunities for selected technological sectors are in this context defined as technological trajectories.

Furthermore, it is here assumed that a fast rate of growth of patenting (i.e. a high rate of growth in patent stock) in a technological sector or class of activity represents an area of strong technological opportunity in the period in question. This is now a standard assumption within patent statistics in general, and the discussion and justification of such measurement is carried out in Chapter 2, section 2.2.

Dynamics

Accordingly, the structural dimensions of the rate and direction of technological activity is a central area of illumination. In relation to the dynamic processes set out in Chapter 1, this chapter operates within relative measures where any stage of a system is relative. That is, a direction of an evolution as well as an opportunity are relative concepts and should be measured and understood relatively. The basic argument is that change in one part of a system affects the position of another. This is one of the essential principles when understanding things a dynamic framework (see Chapter 1, section 1.3 on relative measures). Accordingly, the evolution of the structural dimensions of change in technological opportunities are determined by their relative positions and their distribution.

Furthermore, when investigating development and structural change in technological opportunities focus, in this case, is in on *changes* (i.e. rates or growth) in technological opportunities (as that tells something about the dynamic structures at a given time), rather than its level (as that merely tells something about the accumulated past - important as it may be). That is, as technological sectors grows at different rates (defined by the opportunities which is reflected in the growth rate, see indicators in Chapter 1, section 2), the associated structures change (due to new sorting). This might change the scope and overall configuration of systems of technological development, so that the parameters which determine (i.e. select) the directions of development may change, causing yet new structures and so forth (see Chapter 1, section 1.3 on development, change and emergence).

3.3. TRACING TECHNOLOGICAL EPOCHS AND STRUCTURAL CHANGES IN PATENTING PATTERNS

To get some ideas of the historical trends at the macro-level accumulated patent stocks for the aggregate of all patent classes and for each of the five broad technological groups are displayed in Graphs 3.1 to 3.4. The calculation of the accumulated patent stocks were carried out in Chapter 2, section 2.2.[2]

The graphs certainly suggest changes in technological opportunities over time. It can be seen that the growing opportunities in the science-based sectors (Chemical and Electrical/electronics) have been more or less continuous in the twentieth century, except for some disruptions in the growth rates between 1940 and 1960; while the opportunities in the engineering-based sectors (Mechanical and Transport) as well as the Non-industrial group were weak between the 1930s and up to 1960, over which period those broad technological groups even experienced an actual decline in accumulated patent stock.

The combination of these effects suggest that the appropriate time periods into which to split the analysis are 1920-40 (the Interwar period), 1940-60 (the War/early postwar period) and 1960-90 (the Recent period), as the general picture shows interruptions or breaks in the trends between these periods. The wars referred to in this context, and in this book in general, are World War I and World War II.

However, as economists within the evolutionary tradition emphasise, the macro economy is not simply the aggregate of various micro units but is instead regarded as a complex outcome of micro relationships or interactions, the further analysis will be carried out at a more disaggregated level in order to understand the evolving structures which lies behind the shapes of the aggregate graphs.

---

[2] Stocks were calculated using the perpetual inventory method as in vintage capital models, with an allowance for a depreciation of the separate contribution of each new item of technological knowledge over a thirty year period - the normal assumption for the average life-time of capital, given that new technological knowledge is partly embodied in new equipment or devices, as well as the conventional method used in patent statistics (See Chapter 2, section 2.2. for discussion). Thus the stock in 1919 represents a weighted accumulation of patenting between 1890 and 1919, a thirty year period with weights rising on a linear scale from 1/30 in 1890 to unity (30/30) in 1919, using 'straight line depreciation'.

Graph 3.1. Total accumulated patent stock 1920-1990

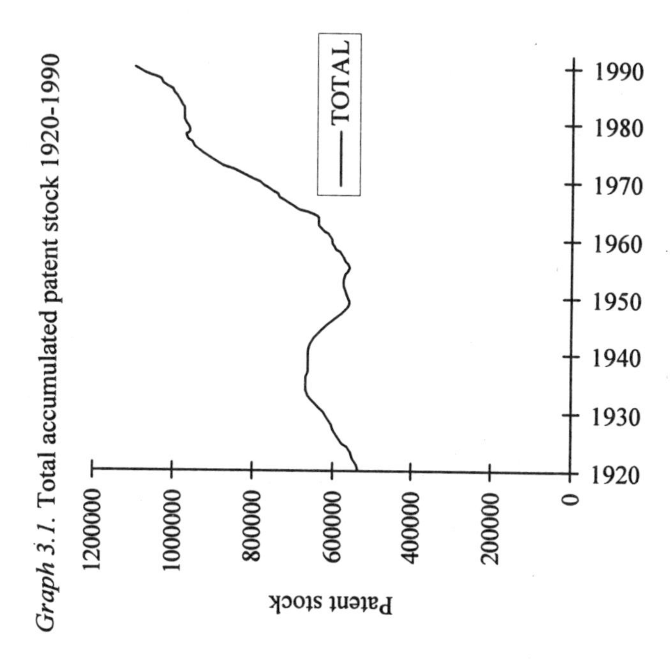

Graph 3.2. Accumulated patent stocks for Chemical and Electrical / Electronic technologies 1920-1990

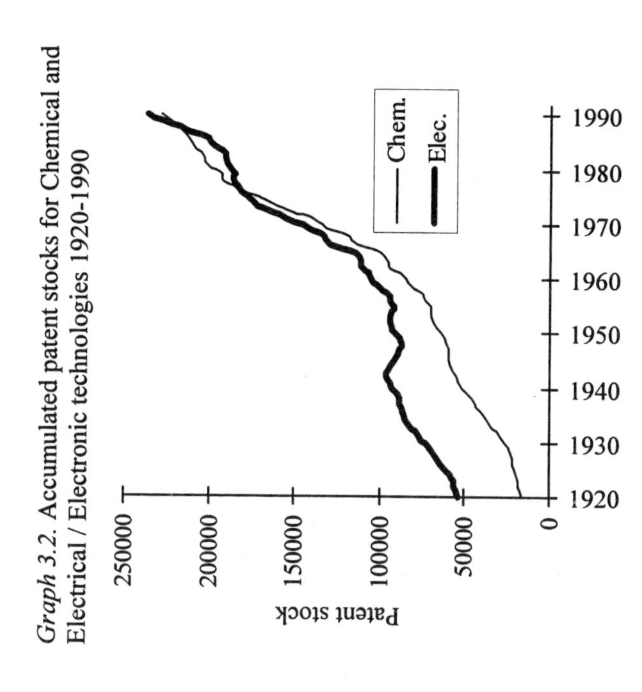

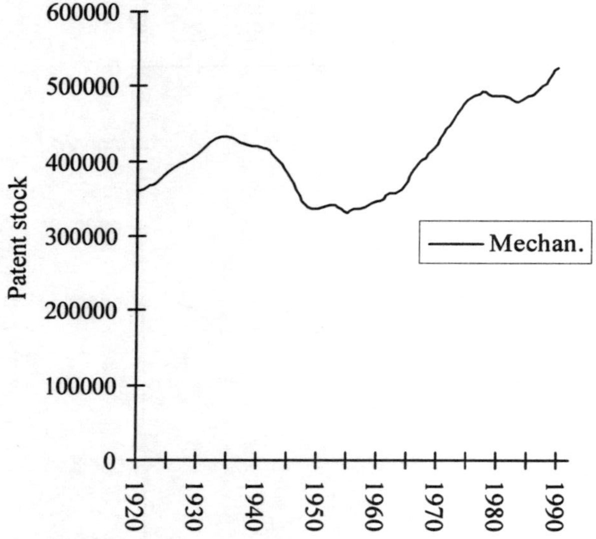

Graph 3.4. Accumulated patent stocks for Transport and Non-industrial technologies 1920-1990

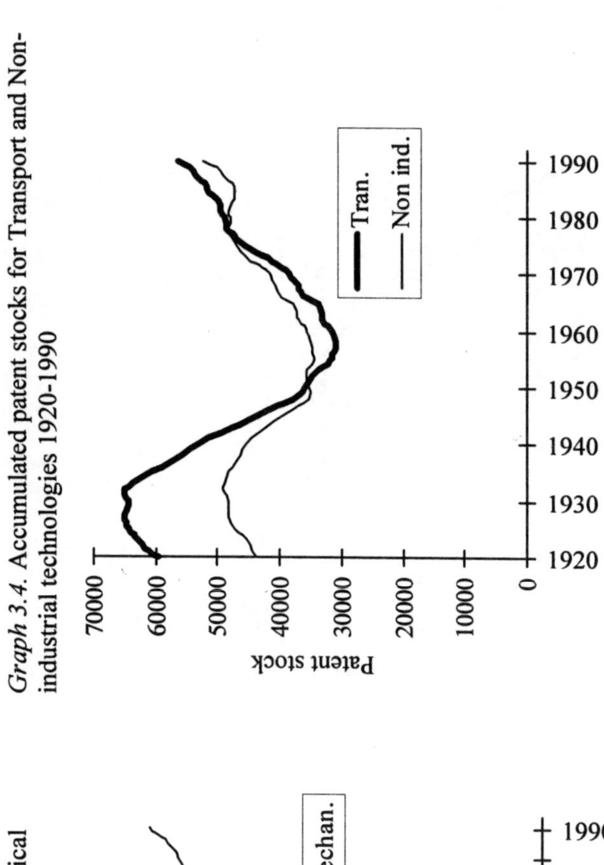

Graph 3.3. Accumulated patent stock for Mechanical technologies 1920-1990

Patent stock

600000
500000
400000
300000
200000
100000
0

1920 1930 1940 1950 1960 1970 1980 1990

Mechan.

When examining the complexities behind the structure of the changing technological opportunities, only patent classes which have at least 10 patents in the beginning and at the end of each of the selected periods have been included in order to avoid problems that can be created with small numbers, which can lead to very high positive or negative growth rates and which represent only random statistical results. Hence, out of the 399 patent classes 369 satisfy the selection criteria, of which the Chemical broad group includes 50 (down by 7), Electrical/electronics accounts for 57 (down by 12), Mechanical for 213 (down by 8), Transport for 21 (no change) and the Non-industrial group for 28 (down by 3).

As shown in Table 3.1, out of the 30 disqualified patent classes, some classes are disqualified because of late arrivals (new technologies coming into importance in absolute terms at some later stage of development, i.e. after 1920); others are disqualified because their technologies have dropped out of absolute importance before the last period(s) of development (i.e. before 1990). Finally, some technological sectors have never been of absolute importance across any single period of development in the twentieth century.

Table 3.1. Inter-group distribution of patent classes

Technological group	Total sectors	Number of technological sectors which:			
		Have ALWAYS been	DROP out	BECOME	Have NEVER been
		of absolute importance during waves of technological development in the twentieth century			
Chemicals	57	50	-	3	4
Electrical/ electronics	69	57	1	8	3
Mechanical	221	213	6	1	1
Transport	21	21	-	-	-
Non-industrial	31	28	-	1	2
TOTAL	399	369	7	13	10

It appears that the engineering-based technologies (Mechanical and Transport) and Non-industrial technologies were mostly already established in some form by 1920, and so had very few new technological sectors arising (2 in total) at a later date - even within the Mechanical group several technological sectors (6 in total) drop out of absolute importance later in the century. On the other hand, many of the science-based technologies (Chemicals and Electrical/electronics) only appeared to be of absolute importance during some later period of technological development in the twentieth century, and had several sectors (11 in total) which first gained in absolute importance after 1920, while Electrical/electronics had only one technological sector dropping out of absolute importance before 1990.

The distribution of the total number of the 369 selected patent classes ranked in accordance to their technological opportunities or growth rates (high, medium or low), whether in absolute or relative terms, and over the three historical periods (i.e. interwar 1920-40; war/early postwar 1940-60; and recent period 1960-90), is presented in Table 3.2.

Table 3.2. Classes within five broad technological groups; ranked in accordance with their technological opportunities or growth rates over three historical periods

Growth rate rankings		Number of sectors			Intra-'broad technological group' distribution in %		
	Period →	1920 - 1940	1940 - 1960	1960 - 1990	1920 - 1940	1940 - 1960	1960 - 1990
HIGH	Chemicals	48	40	32	96.00	80.00	64.00
	Electrical/ electronics	33	31	27	57.89	54.39	47.37
	Mechanical	38	41	48	17.84	19.25	22.54
	Transport	2	1	9	9.52	4.76	42.86
	Non-industrial	2	10	7	7.14	35.71	25.00
	TOTAL	123	123	123	-	-	-
MEDIUM	Chemicals	2	9	14	4.00	18.00	28.00
	Electrical/ electronics	17	16	15	29.82	28.07	26.32
	Mechanical	91	88	78	42.72	41.31	36.62
	Transport	4	4	3	19.05	19.05	14.29
	Non-industrial	9	6	13	32.14	21.43	46.43
	TOTAL	123	123	123	-	-	-
LOW	Chemicals	0	1	4	0.00	2.00	8.00
	Electrical/ electronics	7	10	15	12.28	17.54	26.32
	Mechanical	84	84	87	39.44	39.44	40.85
	Transport	15	16	9	71.43	76.19	42.86
	Non-industrial	17	12	8	60.71	2.86	28.57
	TOTAL	123	123	123	-	-	-

From Table 3.2 it can be observed that the areas of greatest technological opportunities are not concentrated within relatively few areas of presumably related technological fields, but have been increasingly widely dispersed across the five broad technological groups. In this context Chemicals and Electrical/electronics (the science-based sectors), which at the beginning of this century had a relatively high proportion of sectors ranked among the fastest growing, have generally seen a decline in their share of the number of technological sectors ranked among the fastest growing, while Mechanical, Transport (the engineering-based sectors) and Non-industrial, which at the beginning of this century had a low proportion of sectors ranked among the fastest growing, have generally seen an increase in their share of the number of the technological sectors ranked among the fastest growing. However, the greater fluctuations in the relative growth rate rankings of patenting in the Transport and Non-industrial spheres can be partly explained by the relatively small total number of patent classes in those groups in comparison to the other broad technological groups.

Two alternative interpretations concerning changes in the composition of the band with fastest growing technologies (or areas with highest technological opportunities) might be provided, and they are classified into two models, Model A and Model B.

Figure 3.1. Illustrative example of Model A

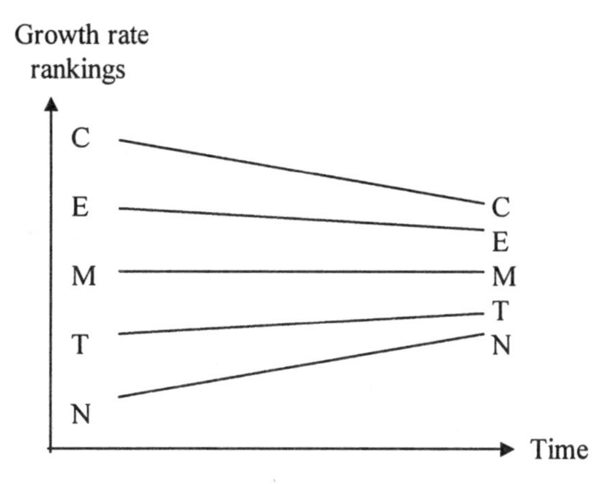

Model A argues that the technological opportunities have become less concentrated or more interrelated due to inter-group convergence in technological opportunities (see Figure 3.1). The alternative explanation comes from Model B, which contends that the technological opportunities have become less concentrated due to intra-group dispersion of technological opportunities (see Figure 3.2).

Figure 3.2. Illustrative example of Model B

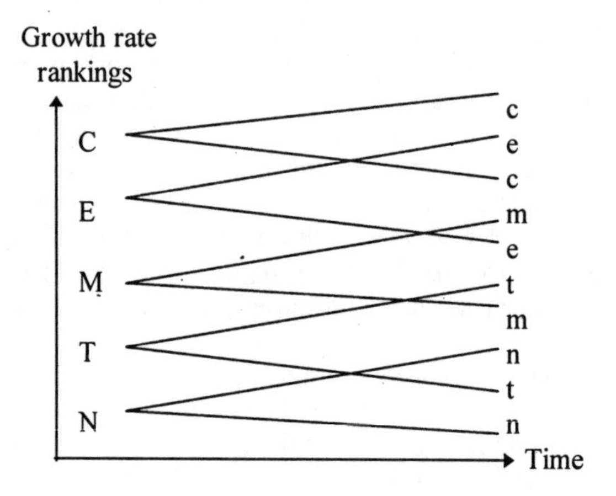

The full description of predictions based on Model A and Model B is described in Figures 3.3 and 3.4, overleaf.

Statistical evidence has been compiled to examine whether either Model A or Model B (or in some cases both of them) account for the trends described above. As the mean growth rates of the five technological groups vary between groups and over time (see Table 3.3), it is difficult to directly compare the degree of concentration and dispersion across classes within groups using the absolute measures provided by the variance or the standard deviation of growth rates, as these depend upon the variate scale. Therefore, this analysis needs a relative measure of the variability of the growth rates across classes in the data which for example is provided by the coefficient of variation, represented by the standard deviation divided by the mean expressed as a percentage: $CV = (\sigma/\mu) * 100\%$.[3]

---

[3] From Table 3.3 it can also be observed that the results concerning convergence versus divergence of growth rates in most cases varies, depending on whether the standard deviation or the coefficient of variation has been used, and that the coefficient of variation is a better measure of concentration due to the great importance of the moving average growth rates over time. Moreover, other measures of concentration, such as the Herfindahl index are also better related to the CV than to the standard deviation. (For a given number of technological sectors there is a strict

Figure 3.3. Predictions based on INTER-group dispersion (or concentration) of technological opportunities*

Cross technological group growth comparisons	Greater INTER-group dispersion between periods	Greater INTER-group convergence between periods
High growth rate ranked groups: (Chemicals and Electrical/electronics)	More in top ranked classes	Less in top ranked classes
Medium and low growth rate ranked groups: (Mechanical, Transport and Non-industrial)	Less in top ranked classes	More in top ranked classes

* *Model A describes how the INTER-group changes in the rankings of patent classes over time might have been associated with INTER-group dispersion (or concentration) in the growth of different broad groups of related technological fields.*

Figure 3.4. Predictions based on INTRA-group dispersion (or concentration) of technological opportunities*

Cross technological sector growth comparisons → Cross technological group growth comparisons	Greater INTRA-group dispersion between periods	Greater INTRA-group convergence between periods
High growth rate ranked groups: (Chemicals and Electrical/electronics)	Less in top ranked classes	More in top ranked classes
Medium and low growth rate ranked groups: (Mechanical, Transport and Non-industrial)	More in top ranked classes	Less in top ranked classes

* *Model B describes how the INTER-group changes in the rankings of patent classes over time might have been associated with INTRA-group dispersion (or concentration) in the growth of different related technological fields collected together in each broad group.*

For Model A the CV has been calculated both (i) using the unweighted average of all the growth rates of all the 369 individual patent classes and (ii) using the unweighted average of each of the five broad groups' unweighted averages); while for Model B the CV has been calculated across individual

relationship between the coefficient of variation and the Herfindahl index; denoting the number of sectors by n and the value of the Herfindahl index by H, the relationship is H = (CV2 + 1)/n (Hart 1971).)

classes' growth rates within each technological group in each period using the unweighted average of all the growth rates within each group. The results of the average growth rates and CVs are displayed in Table 3.3. With respect to the calculation of the CV for Model A, it appears that the results (regarding convergence versus divergence) are not sensitive to the different ways the average of growth rates has been calculated.

Table 3.3. Results concerning the relevance of Model A and Model B in relation to five broad technological groups as well as total*

		1920-1940	1940-1960	1960-1990
Average growth rates: μ (expressed in %)	$\mu_{TOTAL\ 369}$	132.77	15.15	88.98
	$\mu_{TOTAL\ CEMTN}$	163.23	22.61	103.64
	μ_C (Chemicals)	364.48	104.91	200.83
	μ_E (Electrical/ electronics)	405.82	78.93	138.77
	μ_M (Mechanical)	34.36	-13.38	55.54
	μ_T (Transport)	0.65	-43.50	56.14
	μ_N (Non-industrial)	10.81	-13.92	66.94
Model A	(σ_{369})	(405.02)	(114.68)	(147.34)
	$CV_{369} = (\sigma_{369} / \mu_{TOTAL\ 369}) \cdot 100\%$	305.05	756.94	165.58
Measures INTER-group concentration in technological growth rates: (σ) and CVs	$(\sigma_{[\mu_C, \mu_E, \mu_M, \mu_T, \mu_N]})$	(203.48)	(65.09)	(64.41)
	$CV_{CEMTN} = (\sigma_{[\mu_C, \mu_E, \mu_M, \mu_T, \mu_N]} / \mu_{TOTAL\ CEMTN}) \cdot 100\%$	124.66	287.92	62.15
Model B	(σ_C)	(299.76)	(122.21)	(217.07)
	$CV_{Chemicals} = (\sigma_C / \mu_C) \cdot 100\%$	82.24	116.49	108.09
Measures INTRA-group concentration in	(σ_E)	(895.64)	(219.72)	(180.77)
	$CV_{Electrical/Electronics} = (\sigma_E / \mu_E) \cdot 100\%$	220.70	278.36	130.26
Technological growth rates: (σ) and CVs	(σ_M)	(71.68)	(45.40)	(104.13)
	$CV_{Mechanical} = (\sigma_M / \mu_M) \cdot 100\%$	208.60	-339.24	187.50
	(σ_T)	(56.38)	(27.29)	(89.47)
	$CV_{Transport} = (\sigma_T / \mu_T) \cdot 100\%$	8673.84	-62.74	159.37
	(σ_N)	(63.44)	(42.23)	(113.77)
	$CV_{Non\text{-}industrial} = (\sigma_N / \mu_N) \cdot 100\%$	586.66	-303.36	169.96

* *Only the absolute value of the CV is important (hence the positive or negative sign in front of the CV is not relevant).*

By comparing the statistical results of Model A and Model B (by viewing the changes in the CVs displayed in Table 3.3) with the actual general pattern of changes in the composition of the fastest growing classes (as described in Table 3.2) in relation to the model predictions expressed in Figures 3.3 and 3.4, the model can be found which best explains the shifts in the rankings of patent classes.

To use the Chemical broad technological group as an example, we see from Table 3.3 that we have inter-group divergence in technological growth rates between the interwar and the war/early postwar period as the CV for total rises. In accordance with Model A (which deals with inter-group issues - see Figure 3.3) this means that more patent classes within Chemicals would be ranked among the fastest growing. However, from Table 3.2 we see that the Chemical technological group actually declines in share of its patent classes which is ranked among the fastest growing. Hence, model A does not explain the observed empirical evolution.

However, from Table 3.3 it can also be observed that the CV for Chemicals increases between the interwar and the war/early postwar period, reflecting intra-group divergence in growth rates. This would in accordance with Model B (which deals with intra-group issues; see Figures 3.2 and 3.4) cause less chemical technological classes to be ranked among the fastest growing. As we see from Table 3.2 that the broad Chemical technological group actually declines in total number of classes ranked among the fastest growing, the changes in the growth rate rankings of patent classes can be concluded to be explained by Model B in this case.

In that way the models which explain best the complexities behind the evolution of technological opportunities have been found for all broad technological groups, and the results are listed in Table 3.4.

Table 3.4. Best model to explain the evolution of technological opportunities

Technological groups	Period: 1920-40 to 1940-60	Period: 1940-60 to 1960-90
Chemicals	B	A
Electrical/electronics	B	A
Mechanical	B	A
Transport	A and/or B	A and/or B
Non-industrial	(B)	B

In very general terms these results show that Model B dominates between the interwar and the war/early postwar period, while Model A dominates in the more recent period (see Table 3.4). Hence, we find two main technological paradigms governing the last century.

The technological epoch of intra-group divergence in the three biggest broad technological groups - Chemicals, Electrical/electronics and Mechanical - between the first two periods may reflect the formation of specialised engineering and science-based fields, and the period in which the leading sectors of each of these technological groups came to maturity and established the structure for which they are known today. For Transport technologies instead there was intra-group convergence between the interwar and the war/early postwar period; this effect, which reduced the number of high growing Transport classes, was further reinforced by inter-group divergence, as represented by an increased CV across all technological fields and illustrated in Model A. Concerning the Non-industrial technological sectors, neither of the models seems to explain the technological evolution between the interwar and the war/early postwar period. However, if a conclusion has to be drawn, here too model B comes closer to explain the increase in the number of non-industrial patent classes among the faster growing. That is, while the intra-group CV_N falls, there is a tighter bunching of the span of growth rates covered by the 123 sectors in the 'medium' growth range (see Table 3.5). Hence, although the growth rate band among the technological sectors or classes of the non-industrial technologies decreases from [-80.38% to 222.59%] in the interwar period (1920-40) to [-89.86% to 62.78%] in the war/early postwar period (1940-60), which also is indicated by a smaller CV_N, it is still possible for this technological group to increase its number of patent classes ranked in the fast growing section (although lowly ranked within that section), while dropping only by a little the number of patent classes ranked among the low growing technologies.

Table 3.5. Growth rate bands for technological rankings historically

Growth rate rankings	Growth rate band (expressed in %)		
	Interwar (1920 - 1940)	War/early postwar (1940 - 1960)	Recent period (1960 - 1990)
HIGH	71.94 to 4855.98	5.70 to 1436.27	88.34 to 956.51
MEDIUM	7.02 to 71.86	-24.44 to 5.56	12.66 to 88.17
LOW	-80.38 to 6.70	-89.86 to -24.73	-92.14 to 12.63

However, the period between the war/early postwar and up to recent period demonstrates a technological epoch of inter-group technological convergence characterising the evolution of all broad technological groups, except for the very heterogeneous non-industrial technological sectors. This may reflect a new paradigm of formation of broader technological systems as well as development of complex technologies which are offshoots of the incremental nature of technological development. These results can also be interpreted along with von Tunzelmann's notion of growing technological complexity (von Tunzelmann 1995) and Kodama's notion of technology fusion (Kodama 1992). It is suggestive of a historical shift towards more integrated technological systems in recent times through the fusion of diverse and formerly separate branches of technology, which explain the closing of the growth rate gap (illustrated in Tables 3.2 and 3.3) between the science-based technologies (Chemicals and Electrical/electronics) and the engineering-based technologies (Mechanical and Transport) as well as the Non-industrial technologies. This evidence concerning closer connections between the principal technological families also supports the study by Patel and Pavitt (1994). In other words, technological development has become increasingly interrelated and complementary rather than independent and distinct.

That the evolution of Transport technologies up to recent times is, as the only broad technological group, characterised by Model B, showing dispersion of intra-group growth rates, might be related to the revival of certain transport technologies in recent times, as seen in Graph 3.4.

However, what is interesting is not only the *overall* changes of structures, but to examine the nature by which the technological opportunities have evolved. Section 3.4. will investigate the nature by which the technological opportunities have changed their relative positions and changed in their distribution, whereas section 3.5 will identify and examine individual trajectories of technological opportunities.

3.4. THE NATURE OF EVOLVING TECHNOLOGICAL OPPORTUNITIES

The above results concerning *overall* structural changes in patenting patterns will now be studied in more detail, with reference to the nature of the changing compositions of specific technological sectors. The nature by which the technological opportunities have changed their relative positions and changed in their distribution will be the unit of analysis.

Within post-Schumpeterian approaches to evolutionary institutional economics it is often argued that the structure of technology, and therefore also technological opportunities (as well as most other variables, such as e.g. technological capabilities, see Chapter 7), changes only gradually or incrementally over time (see e.g. Rosenberg 1976),[4] as opposed to radically. However, the direction(s) of these incremental changes is often unclear; such as e.g. the extent that gradual or incremental changes cause convergence or divergence of technological activity.

These issues concerning (i) high mobility or incremental nature of changes as well as (ii) whether such change causes convergence or divergence, can be measured empirically/statistically, by examining the changing composition of technological growth rates across different historical periods.

Such analysis is better done at the intra-group level due to the great difference in absolute size (i.e. number of technological fields or classifications) of the broad technological groups, which would eliminate the smaller broad technological groups' overall contribution to the results. This would display only Electrical/electronics and Mechanical major historical regimes rather than evolving structures. The analysis is based on the 369 eligible patent classes selected in the previous section. Again based on work in the previous section, appropriate time periods into which to split such analysis are 1920-40 (the interwar period), 1940-60 (the war/early postwar period) and 1960-90 (the recent period).

The Model

Though a statistical analysis, cross-section changes in the pattern of the patent growth rates are decomposed into a 'mobility effect' and a 'regression effect'. The 'mobility effect' tests for the degree of mobility in the growth rate rankings of patent sectors: big mobility or change in growth rate rankings (indicating destruction or radical changes of existing structures) versus small change in rankings (indicating incrementalness) versus no change. The 'regression effect' tests for the direction of the evolution: convergence vs. divergence vs. no changes. The statistical principle is commonly known as a 'Galtonian regression' (See Hart 1971, 1994, Cantwell and Andersen 1996. See also Chapter 8 where the model is applied in relation to the evolution of corporate technological leadership).

In this framework, the regression coefficient β measures the intra-group changes in the distributions of patent classes' relative growth rates over time (i.e. whether the intra-group growth rates tend to move closer to or further away from the mean). The magnitude of the 'regression effect' is measured by $(1-\beta)$, as for $\beta=1$ (or $1-\beta=0$) there is on average no intra-group convergence or divergence of growth rates.

---

[4] This argument has also been applied from biological analogies to economics, although the limitations of such analogies has been recognised Hodgson (1993).

As mentioned above, the other feature arising from the regression analysis is the simple test of the extent of mobility or fluctuations in growth rates across patent classes (i.e. whether the intra-group variations in growth rates indicate patterns of stability or fluctuations in rate of technological activity or opportunities across sectors over time). In this framework, the correlation coefficient, ρ, is a measure of degree of mobility or fluctuations of patent classes' growth rates over time. The magnitude of the 'mobility effect' is measured by $(1-\rho)$, as for $\rho=1$ (or $(1-\rho=0)$ there is no mobility.

The Estimations

Two simple cross-section regressions of the technological sectors' growth rates expanding over three broad periods (from the interwar t-2 to the war/early postwar t-1; and from the war/early postwar t-1 to the recent period t) are carried out. The analysis is done separately for each broad technological group, as mentioned above.

However, it is found that it is better to express the growth rates in logarithmic form (since the distribution of the size and hence growth is closer to a log normal than a normal distribution):

(i) $\text{Log}(GT_i + 1) = \text{Log}\, P_{i(t)} - \text{Log}\, P_{i(t-1)}$

where GT denotes the rate of growth, and P denotes the patent stock of sector i in time t (or t-1).

Hence, the changing composition of the technological opportunities can then be statistically tested using following cross-section regressions:

Regression 1.

Between the interwar (1920-40) and the war/early postwar period (1940-60)

(ii) $[\text{Log}(GT + 1)]_{i(t-1)} = \alpha + \beta\, [\text{Log}(GT + 1)]_{i(t-2)} + \varepsilon_{(t-1)}$

where [Log (GT + 1)] refers to the logarithm of rate of growth in patent stock i over the time period in question.

Regression 2.

Between the war/early postwar (1940-60) and the recent period (1960-90)

(iii) $[\text{Log}(GT + 1)]_{i(t)} = \alpha + \beta\, [\text{Log}(GT + 1)]_{i(t-1)} + \varepsilon_{(t)}$

where [Log (GT + 1)] refers to the logarithm of rate of growth in patent stock i over the time period in question.

The Results

The results are displayed in Table 3.6 and 3.7. From Tables 3.6 and 3.7 it can be observed that especially the relatively newly established science-based groups (Chemicals and Electrical/electronics) have each experienced a strong regression towards the mean throughout the century, reflecting an integration of historical (or relatively old) and new fields of development. In order words, the knowledge embodied in the old regime, operating between the interwar and war/early postwar period, is also embodied in the new regime which has been governing the period up to recent time.

Moreover, Chemicals within the science-based sectors have experienced high mobility or disruptions in the compositions of the intra-group growth rates, but this only in recent times. Electrical/electronics has experienced moderate mobility in the composition of new technological opportunities throughout the century. This 'mobility effect' may suggest that the relatively newly established chemical industry is still going through some degree of 'search and selection' process of development.

Table 3.6. The extent of continuity (convergence or divergence) in the composition of technological opportunities within broad technological groups over time

Technolo-gical broad group	Numbers of sectors	Period	REGRESSION EFFECT		
			$\hat{\beta}$	$(1-\hat{\beta})$	$t_{\beta 1}$
Chemicals	50	Regression 1	0.450	0.550	-4.632***
	50	Regression 2	0.119	0.881	-5.371***
Electrical / electronics	57	Regression 1	0.272	0.728	-8.581***
	57	Regression 2	0.363	0.637	-4.263***
Mechanical	213	Regression 1	0.298	0.702	-12.701***
	213	Regression 2	0.637	0.353	-3.682***
Transport	21	Regression 1	0.727	0.273	-1.343
	21	Regression 2	0.618	0.382	-1.386*
Non-industrial	28	Regression 1	0.506	0.494	-2.576**
	28	Regression 2	-0.047	1.047	-4.928***

Statistically significant at the 1% level (***), at the 5% level (**) and at the 10% level (*).

Concerning the older and more mature engineering-based sectors (Mechanical and Transport), the 'regression effect' has decreased significantly concerning Mechanical technologies, while this effect for Transport has been very low throughout the century and was not even statistically significant in the shift up to the war/postwar period. Concerning the 'mobility effect', it has not been strong at all for the engineering-based sectors. These findings seem to indicate that for the older and more mature engineering-based sectors less intra-group movements and integration of technological fields, as well as no disruptions in the evolution and compositions of technological opportunities, is to be expected in general.

Table 3.7. The extent of continuity (creative destruction or accumulative incrementalness) in the composition of technological opportunities within broad technological groups over time

Technological broad group	Numbers of sectors	Period	MOBILITY EFFECT		
			$\hat{\rho}$	$(1-\hat{\rho})$	t_{B0}
Chemicals	50	Regression 1	0.480	0.520	3.786***
	50	Regression 2	0.105	0.895	0.728
Electrical / electronics	57	Regression 1	0.397	0.603	3.204***
	57	Regression 2	0.312	0.688	2.431**
Mechanical	213	Regression 1	0.349	0.651	5.411***
	213	Regression 2	0.407	0.593	6.466***
Transport	21	Regression 1	0.634	0.366	3.572***
	21	Regression 2	0.457	0.543	2.241**
Non-industrial	28	Regression 1	0.459	0.541	2.637**
	28	Regression 2	0.045	0.955	-0.220

Statistically significant at the 1% level (***), at the 5% level (**) and at the 10% level (*).

Concerning Non-industrial technologies, the 'regression effect' has been strong and significant only in recent times; even to the extent that formerly slower growing technologies have been overtaking the formerly faster growing. Within the Non-industrial technologies there has also been an increasing 'mobility effect' in recent times, although not statistically significant. However, it here ought to be mentioned again that this broad technological group is a residual group not based on technological interrelatedness, and that this overtaking of formerly slow growing technologies may just reflect the falling back of war based technologies, some of which are included in the Non-industrial group and which showed high opportunities in the war/early postwar period.

These overall results suggest that the evolution of twentieth-century opportunities indicate patterns of uniform movements across different periods of technological development, rather than being marked by major disruptions and random mobility. Hence, the composition of the technological opportunities, or sectoral growth rate ranking positions, does not tend to fluctuate across different historical waves of technological development. Together with a significant 'regression effect' (although not very strong for the more mature engineering-based broad groups), this reflects that knowledge embodied in both older and newer fields of technological opportunities contribute to the development of a new technological paradigm; this evidence supports the now common view of creative incremental accumulation being dominant in recent times rather than radical shifts of 'creative destruction', and that new paradigms generally do not destroy old ones but complement and extend them (Pavitt 1986; Patel and Pavitt 1994).

Overall, this evidence supports the now common view of 'creative, incremental accumulation' with interconnected technological evolution. In this framework, technologies previously following isolated channels of development (as identified in section 3.3) are brought together, in a fashion in which knowledge from old systems is generally not destroyed in newer systems, but complemented and extended. This is what is causing blurring of technological opportunities in recent times. This is consistent with the evidence found in Pavitt (1986) and Andersen and Walsh (2000).

This 'creative incremental accumulation' in technological development does however not exclude other variables (such as institutions, corporate structures, the systems themselves etc.) from experiencing creative destruction across technological epochs[5].

---

[5] See Chapter 6 on technology dynamics in which the co-evolution of technology and industry structures (as well as the consequences of such) are examined. See also Chapter 7, which documents how the structure of the technological profiles of firms becomes eroded across technological regimes.

3.5. REVEALED TECHNOLOGICAL TRAJECTORIES

Whereas the analysis so far has been based upon structures and the nature in which such structures changers, this section aims to identify the individual trajectories of technological opportunities. The scope of this section is that, given the Schumpeterian institutional approach taken in this book, it is believed that the broad technological groups do not follow any general trajectory of changing technological opportunities (i.e. is gathered in only one trajectory type). However, it is expected that some trajectories are more likely to be followed rather than others due to the 'instituted' nature of technological development, as defined by the existing technological paradigm and structure of socio-economic competence (Nelson and Winter 1977, Dosi 1982, 1988), see section 3.2. So rather than identifying randomly and unstructured the evolution paths of the selected 369 technological sectors (or patent classes), it seems more appropriate to examine to what extent a technological trend is typical.

Thus, the next step of the analysis is to examine and identify how the underlying patterns of trajectories of technological opportunities, with respect to the individual 369 technological sectors, have evolved over time. This is done by deriving typical technological trajectories and selecting those of greatest historical technological importance.

The Model

Different possible trajectory types for the growth performance of patenting in each class can be identified using a general framework which has been developed for this purpose (see Figure 3.5). In order to measure the evolution of trajectories (understanding that the direction of an evolution as well as an opportunity are relative concepts), all Chemical, Electrical/electronics, Mechanical, Transport and Non-industrial technological opportunities (or growth rates) have been ranked within a framework (as in Table 3.2), which sorts all 369 eligible technological sectors or patent classes into nine groups, derived by ranking them into three bands in accordance to their growth rates within each of three broad historical periods (interwar 1920-40, war/early postwar 1940-60, and recent period 1960-90). The growth rate ranking is done across all the broad technological groups, as in Table 3.2.

This scheme can then be used as a framework to reveal many different trajectory types showing alternative paths or ways in which the 369 eligible individual technological sectors may have changed their relative opportunities over time. Seventeen possible alternative trajectory types can be traced, when sorted in accordance with their initial starting point (whether they start high, medium or low growth rate ranked), and if they evolve in a linear or non-linear fashion. As appears below, there are also different types of linear and non-linear evolution paths.

Patent classes or technological sectors which follow a linear horizontal trajectory [1 (111); 2 (222); 3 (333)] reflect those technologies whose technological opportunities remain constant historically. Also, there are quadratic-like horizontal types of trajectories. If they are first declining - but then recovering [6 (121, 131); 7 (232)], these seem to be technologies with dropping opportunities in the war/early postwar period. However if they are first growing but then falling back to their initial starting point [12 (212); 13 (323, 313)] this indicates technologies with indeed great opportunities in the war/early postwar period.

Figure 3.5. Framework for identification of trajectories of technological opportunities

Relative growth rate ranking position	Historical time trend		
	Interwar (1920 - 1940)	War/early postwar (1940 - 1960)	Recent period (1960 - 1990)
HIGH	1	1	1
MEDIUM	2	2	2
LOW	3	3	3

Finally there are historical declining and historical growing trajectories. Declining trajectories can be linear 5 (123), non-linear convex-like [10 (122, 133, 132); 11 (233)] or non-linear concave-like [16 (112, 113); 17 (223, 213)], and they all indicate historically falling opportunities. However, technological sectors performing growing trajectories, which can also be linear 4 (321), non-linear convex-like [8 (221, 231); 9 (332, 331)] or non-linear concave-like [14 (211); 15 (311, 322, 312)] indicate an historical growth in new opportunities.

Such an analysis of identifying typical technological trajectories will be carried out at the broad technology group level. The reason for carrying such an analysis out at the group level is mainly to adjust for size (i.e. number of technological fields or classifications), so that related technological sectors within the larger groups such as Electrical/electronics and Mechanical do not bias the results concerning which trajectories have been typical and dominant. This would display only Electrical/electronics and Mechanical major historical regimes. Hence, in accordance with section 3.2. this analysis takes into account that what is interesting is not only the overall weight or significance of a trajectory for a technological group, but also a trajectory's relative typicality for this technology group relative to other broad groups (such as Electrical/electronics, Mechanical, Transport and Non-industrial), which is also another way of measuring a group's relative contribution to specific paths of development.

The Estimations

A trajectory's typicality for a technological group relative to other trajectory types and in comparison to those that characterise other groups, can then be measured by an index termed the Revealed Technological Trajectory (RTT), first developed in Andersen (1998). The RTT index is a relative measure, and it can be compared to the Revealed Technological Advantage index (RTA) (see Chapter 7 and 8) as well as the Revealed Comparative Advantage Index (RCA) which is commonly used in trade theory.

The value of the RTT index measures a broad technological group's share of technological sectors (or patent classes) following a particular technological trajectory type, relative to the overall share of all patent classes following this particular trajectory type. Hence, denoting by F_{Tj} the number of technological fields or sectors which follow technological trend T for a particular broad technological group j, the RTT index for each trend type in the broad technological group in question can then be defined as in equation (iv).

$$\text{(iv)} \qquad RTT_{Tj} = \frac{F_{Tj}/\sum_T F_{Tj}}{\sum_j F_{Tj}/\sum_T \sum_j F_{Tj}}$$

where $F_{Tj}/\sum_T F_{Tj} \geq 0.1$ (10%)

The index varies around unity such that when RTT > 1, the trajectory type in question is relatively typical for the technological group in question; hence the group has a relatively massive contribution to this particular technological development path or trajectory. However, RTT < 1 indicates that the trajectory in question is relatively uncommon for the technological group in question.

The RTT index of each of the 17 technological trajectory types has been calculated for all broad technological groups. However, only RTT results of trajectories which each accounts for at least 10% of its technological group's patent classes will be considered, to make sure that only trajectories which are of great overall historical importance for the broad technological groups are revealed.

The Results

The typical technological trajectories of great overall importance for each broad technological group (i.e. cases in which RTT>1 and which have an overall technological broad group patent share of about at least 10%) are displayed in Table 3.8.

Table 3.8 shows that Electrical/electronics, Mechanical and Non-industrial all have five different technological trajectories which are of great importance for the groups and typical relative to evolution paths in other groups, while Transport has four. However, Chemical technologies seem to be gathered in only three typical technological trajectories of great importance.

These findings certainly suggest that it would be an oversimplification to draw any generalised conclusions concerning typical and historical important technological trajectories at the aggregate group level, as the different technological sectors belonging to each of the technological broad groups show a quite complex intra-group pattern in their technological evolution with alternative typical technological trajectories going in totally different directions across different periods of technological development. Hence, all broad groups contribute to various alternative historical technological important paths of development. The typical trajectories of great importance for each broad technological group, as presented in Table 3.8, will now be explained.

Table 3.8. Typical technological trajectories of great importance

Technological broad groups	Number of important trajectories where RTT > 1	Trajectory types*
Chemicals	3	1, 6, 16
Electrical/electronics	5	1, 6, 14, 16, 17
Mechanical	5	2, 3, 9, 15, 17
Transport	4	3, 4, 8, 9
Non-industrial	5	8, 9, 12, 13, 15

*See text for Figure 3.5.

Whereas the patent names at class level will be integrated in the text, the technological groups in which the classes (or a subdivision of a class) are grouped (into 56 technological groupings in total - see Chapter 2) will be presented within 'apostrophes'.

Concerning Chemicals, as many as 44 out of the group's 50 eligible classes (i.e. 88%) are gathered in three typical technological trajectories showing historical important paths of development. In this context, technological sectors documenting continuous opportunities throughout (trajectory 1) include the overall set of subclasses belonging to 'Agriculture chemicals'. Also the overall set of subclasses belonging to 'Photographic chemistry', and most patent classes within 'Synthetic resins and fibres' (including rubber, plastics and adhesives from the polymer industry), as well as most classes belonging to

'Pharmaceuticals and biotechnology' are gathered in this trajectory of continuous opportunities throughout; and so is a great part of the patent classes belonging to 'Chemical processes' and 'Other organic compounds'. Another typical technological trajectory within Chemicals indicates sectors which drop out of the high opportunity ranking position in the war/early postwar period but then recover in later times (trajectory 6). This trajectory is most typical for patent classes belonging to chemistry of 'Inorganic compounds' as well as textile chemicals including 'Bleaching and dying'. In fact, the overall set of subclasses within those latter mentioned groups follow this trajectory type. Finally, the last typical trajectory within Chemicals indicates classes which have enjoyed great technological opportunities until 1960 but then drop out in recent times (trajectory 16). This group includes half of the subclasses within 'Distillation processes' (the other half dropped out of high opportunities already after 1940), as well as most classes within 'Coal and petroleum products' which especially peaked in development in the period ranging from the interwar and up to and including the early postwar period. Also about half of the sectors within organic compounds are gathered in this trajectory type.

The Electrical/electronics group's five typical technological trajectories of great importance for its technological group include 36 out of the sector's total of 57 eligible patent classes (i.e. 63%) selected for this analysis. The technological sectors' contribution to the specific paths of development indicated in the five typical trajectory types are in most cases spread across most of the 56 technological groupings belonging to Electrical/electronics. However, by investigating the specific classes within each of the groupings in relation to the five typical trajectories, and if any overall evolution trends have to be drawn, the five typical trajectories divide between sectors which have been of continuous importance throughout the century (trajectory 1), such as classes within 'Telecommunications', 'Office equipment and data processing systems' as well as 'Semiconductors', and other sectors which were just enjoying higher or greater opportunities at some technological epoch(s) of this century. Sectors with high opportunities in the beginning of the century, or growing in opportunities up to recent period but have since fallen in importance (trajectories 16 and 17), are typical sectors within the electrical equipment industry and technologies of 'Electrical devices and systems'. Sectors which have risen in importance (such as trajectory 14) are typically sectors within 'Communication systems' as well as 'Other electronic devices' including optics, LASER and space-technology. Finally, sectors which show typical trajectories which indicate a drop in opportunities during the war/early postwar period (trajectory 6) cannot be generalised but spread across a broad range of different technological fields. However, among the classes following this trajectory about half of the classes within 'Image and sounds equipment' as well as 'Photographic equipment' seem to follow this trajectory.

Within Mechanical only 109 out of the eligible 213 sectors in total (i.e. 51%) contribute to development path which are typical as well as important for the broad group. Hence, this shows that we are dealing with a very technological heterogeneous sector, which is hard to group into any particular or typical paths of development. Furthermore, within Mechanical only a few of the typical technological trajectories, and only some of the technological sectors within them, reflect technologies which have shown high technological opportunities at any point of time in the twentieth century; although Mechanical technologies as a group as a whole have been of great absolute importance in terms of accumulated technological size (or patent stock, see Graphs 3.1 to 3.4). Also here, as within Electrical/electronics, the technological sectors' contribution to the specific paths of development indicated in five typical trajectory types are not clustered in certain broader technological categories but spread across a range of 56 technological groupings. Sectors belonging to those few trajectories which have shown increasing technological opportunities (trajectories 9 and 15) are e.g. technologies belonging to 'Miscellaneous metal products', 'Material handling equipment', 'Agriculture equipment', 'Food, drink and tobacco equipment', 'Other general industrial equipment', 'Power plants' etc. However, sectors which started with great opportunities but then indicated decreasing opportunities (trajectory 17) are e.g. 'Other specialised machinery' (wrapping, brushing, coating etc.), metal working equipment, stone working, paper making apparatus etc.

The Transport group's four typical and overall important technological trajectories presented in Table 3.8 include as many as 17 out of the sector's 21 eligible patent classes (81%). A high proportion of Transport technologies seems to have been gathered in a trajectory with sectors continuously lowly ranked throughout this century (trajectory 3). These are particular technological sectors belonging to 'Railway and railway equipment' or technologies concerning wheels and axles within 'Transport equipment'. However with regard to the other typical and overall important trajectories within Transport, trajectories seem to show a rising tendency and an increase in technological opportunities (trajectories 4, 8 and 9) rather than a fall. Technological fields belonging to such growing trajectories are the overall set of subclasses within 'Internal combustion engines', the full set of classes belonging to 'Motor vehicles', as well as a great part of the technologies within 'Ships and marine propulsion'.

Within Non-industrial 19 out of the eligible 28 sectors in total (i.e. 68%) contribute to development of notable or important trajectories which indicate typical development paths for the broad group. Among the Non-industrial technologies two out of the five typical technological trajectories indicate sectors which are increasing in technological opportunities historically (trajectories 8 and 9); while other typical trajectories show especially greater opportunities in the war/early postwar period (trajectories 12, 13) after which they fall back to their initial starting point. However, also several sectors within the trajectory type 15, which show a drop in opportunities in recent times, actually peaked among the areas of greatest opportunities in the war/early postwar period. Sectors which are rising in technological opportunities are exclusively technologies within 'Other manufacturing and non-industrial' such as e.g. bridges, plant and animal husbandry, amusement devices and games etc.; while firearms, ammunition, explosive-charge making, ordnance as well as education and demonstration etc., (also part of 'Other manufacturing and non-industrial') are among the sectors which peaked in the war/early postwar period, and so did some of the technologies of knot and knot tying as well as apparel within 'Textiles, clothing and leather'.

From here it can be concluded that typical technological trajectories governing different subgroups or sectors within each broad technological group explain technological evolution better than aggregate technological trajectories of broad groups as a whole, as all broad groups contribute to various alternative paths of development as captured by the RTT index.

It is here believed that sectors within each broad technological group, which contribute to a specific path or trajectory of development which is typical and important for that group in question, may be related to families of interrelated technologies. Moreover, trajectories of technological opportunities which indicate similar development paths, or show similar opportunities across broad groups, may even be related to broader interrelated technological families. Yet, whether technological sectors grouped together within each of the typical technological trajectories, and whether typical trajectories within similar development paths across broad groups, indicate interrelated technologies which could be interpreted as technological families require a much more elaborate analysis which is outside the aim of this Chapter.

However, based on another study (Andersen 1997), describing a century of technological opportunities in which the development paths of related technologies were investigated qualitatively, the results concerning such a relationship are very promising.

An example from that study will now be given in relation to the typical trajectories identified above. This example also documents how, after more isolated channels of development, the technological source sectors and diffusion sectors have become less focused and more complex over time; a result or belief which was also supported quantitatively in Section 3.3 concerning structural changes in patenting patterns.

That chemical engineering in the interwar period went through an epochal shift from coal-based to petroleum-based feedstocks pushed a whole oil-based type of a paradigm up to and including the early postwar period, based on coal and petroleum products, distillation processes and development of new and better fuels for engines (trajectory 16); as well as a new range of materials from polymers (e.g. synthetic rubber, plastics, adhesives, man-made fibres (e.g. nylon), teflon and many more) and other organic compounds whose opportunities have continued up to the present (trajectory 1). Similarly, the electrification and the development of electrical devices (trajectories 16 and 17) within the Electrical/electronics broad group up to and including the Early postwar period have probably also been one of the most consequential technological changes. However, the more recent development of a new kind of paradigm of complex electronic based technologies (as e.g. electronic devices and related instruments including optics) (trajectory 14), as well as the continuous opportunities in technologies related to information and communication) (trajectory 1), would simply not have been possible without inventions and innovations within organic chemistry including the synthetic polymer industry. Freeman (1963) and Day (1990) argue how the polymer industry made electrical and electronic engineering manageable, and how it had essential applications in developing good electric insulators and advancing electrical and electronic engineering. Likewise, Chandler (1990, p.217-221) emphasises how research in large companies, such as General Electric, became in direct competition with companies from the polymer industry, such as Du Pont, from their research on insulation for wire and moulding of carbon light bulbs etc. Other studies on polymide applications in electronics include Grupp and Schmoch (1992) and van Vianen and van Raan (1992) who studied the crossroads in polymide chemistry and electronics focusing especially on LASER technology applied in medicine. This is just one example out of many concerning how technological families may have evolved and how technologies have become more complex. The evolution of possible 'waves' of interrelated innovation systems of generic technologies is examined wholly in Chapter 5, and its co-evolution with industry structures is examined in Chapter 6, which deals with technological systems and industry dynamics.

Thus, it can be argued that the quantitative results presented here, concerning typical technological trajectories of areas of greatest technological opportunities for each broad technological group, are not a pure statistical or random phenomenon, but match up quite nicely with what has been suggested in the history of technology literature and other case studies on technologies chosen for their particularly important contributions to development.

3.6. CONCLUSION

It can be concluded that the areas of greatest technological opportunities during the century ranging from 1890 to 1990 are not strongly concentrated within relatively few areas of related technological fields but have been increasingly widely dispersed across broad technological groups over different epochs of technological development; and that the evolution of the last century's technology can be divided into two major technological regimes or paradigms.

Whereas the first technological regime extending from the opening of this century until the war/early postwar period was characterised by intra-group technological diversification and the formation of a structure of specialised engineering and science-based fields, the one that has followed through to recent time (in which the gap between the science-based (Chemicals and Electrical/electronics) technologies on the one hand, and the engineering-based (Mechanical and Transport) technologies and Non-industrial technologies on the other hand has been widening less quickly), is suggestive of an historical shift towards more integrated technological systems through the fusion of diverse and formerly separate branches of technology. The new paradigm governing the evolution paths of trajectories of technological opportunities builds to a greater extent on inter-group complementary and interrelatedness rather than more isolated individual channels of development.

When exploring the nature by which technological evolution has increasingly been converted or channelled into wider-ranging and more complex technological systems, we saw how new systems are offshoots of a creative incremental technological development process in a variety of areas (as opposed to creative destruction), in which knowledge embodied in old technological fields within old systems is integrated in newer systems. This is witnessed by the fact that knowledge embodied in old paradigms is generally not destroyed but complemented and extended in new ones.

Finally, evidence shows how all broad technological groups' relative contribution to specific technological paths or trajectories have contributed to several alternative directions of opportunity development. This suggests that it is inappropriate to draw any 'general' conclusions concerning the specific contribution to the technological development of selected aggregate broad technological groups. This also demonstrates how revealed technological trajectories explain technological evolution better than the conventional aggregate measures which give an illusory picture.

Thus, this chapter certainly supports the view that technology changes and trajectories evolve in an incremental, accumulative and path-dependent fashion, and that some trajectories are more likely to be followed than others.

4. The Hunt for *S*-Shaped Growth Paths in Trajectories of Technological Innovation[1]

Abstract

Since the works of the business cycle theorists in the 1930s, no attempts have been made to study empirically the long term evolutionary paths of individual technologies using long time series. This chapter is an empirical exploration and confirmation of the now almost uniform image or metaphor of the way technology develops; that it follows an *S*-shaped growth path which is commonly associated with a similar shaped diffusion function of entrepreneurial activity. The chapter also confirms the diversity of technology dynamics and explores how technological cycle takeoffs appear to be clustered within certain historical epochs. The results have implications for our understanding of the evolutionary paths of individual technologies, and the evolution of technological systems or waves of innovation.

By use of computational statistics, logistic growth functions are fitted to US patent stocks, 1920-90, at a detailed level of aggregation, including Chemical, Electrical/electronic, Mechanical, Transport and Non-industrial technologies. Some practical considerations when developing an empirically testable model of innovation cycles are addressed in the chapter.

4.1. INTRODUCTION: THINKING IN *S*-SHAPES

Following on Schumpeter's (1939) contribution to business cycle theory, which deals with the association between the distribution of entrepreneurial ability and cyclical movements in the rate of innovation (Kuznets 1940, p.114), it is now commonly argued among Schumpeterian economists that the development and diffusion phase of a selected variable generally follows a sigmoid curve (popularly termed an *S*-curve). Such studies have, for example, covered the

---

[1] This chapter is a revised version of Andersen (1999a)

areas of population growth, technology growth, diffusion of innovations and market shares, and much more.

Within economics the most prominent literature on *S*-curves concern the product life cycle literature (e.g. Kuznets 1930a, Burns 1934, Vernon 1966, Utterback and Suarez 1993, Utterback 1996, Klepper and Simon 1997), or are related to the diffusion process of technologies among firms (e.g. Mansfield 1961, Silverberg, Dosi and Orsenigo 1988, Cantwell and Andersen 1996). However, the 'product life cycle'-based literature has been concerned with the characteristics of products, technologies and industries along associated life cycles, while treating the cycles themselves in a very informal way. The diffusion literature within evolutionary economics, likewise, seems to have just assumed that the diffusion curves were related to associated *S*-shaped development or growth curves. Related literature within the same tradition has focused on diffusion processes of technologies within societies (e.g. the extensive amount of literature on innovation systems, Lundvall 1992, Nelson 1992), or focused on cross-sectional analysis of the structural dimensions of the way technologies develop in an increasingly interrelated complex fashion (see e.g. Kodama 1986, 1992, von Tunzelmann 1995, Andersen 1998). Finally, literature based on 'long waves' in economic life (Kondratiev 1925) has also been associated with *S*-curves. This has mostly been in relation to economic booms or crises or on identifying clusters of waves of innovations (e.g. Mensch 1975, Freeman, Clark and Soete 1982, Solomou 1986, Kleinknecht 1990, see also Chapter 5), rather than identifying to what extent the individual development patterns themselves are *S*-shaped.

Hence, since the empirical works by the business cycle theorists in the 1930s (Kuznets 1930a, Merton 1935), no recent efforts have been made to study empirically the long term evolution paths of individual technologies starting with long time series.[2] Given the lack of empirical evidence on *S*-shaped growth patters, the *S*-curve is still merely an image or metaphor of the way technology develops, as argued by Lindquist (1994). Therefore, the key issue addressed in this chapter is an empirical test of the theoretical use of the *S*-curve in relation to the *S*-shaped innovation growth cycle of technological development.[3] Thus, this

---

[2] Griliches (1957) studied output growth related to the diffusion of hybrid seed corn among farmers in the US.

[3] The aim of this chapter is *not* a theoretical one discussing why it is believed that technology develops not in a smooth line but in cycles. These issues have been dealt with plentifully elsewhere in the business cycle literature on diffusion processes of innovations, 'entrepreneurial swarming' and processes of retardation, as well as in the literature on economic booms and crises or 'clustering of innovations'; see e.g. Schumpeter 1939, Kuznets 1930a, 1940, Kondratiev 1925, Freeman, Clark and Soete 1982, Solomou 1986, and Kleinknecht 1990.

study can be seen as a recent contribution to an analysis of the long-term evolutionary growth paths of individual technologies. This chapter also stresses some practical considerations concerning the development of an appropriate empirically testable model of *S*-shaped growth. Still, despite the great use of *S*-curves or *S*-shaped development patterns, up to the present there is no dominant theoretical framework relating to *S*-curves, or any conclusive associated statistical tests of such. Hence conceptual, technical and statistical issues are given importance in this chapter.

The diversity of technology dynamics is also commonly recognised in post Schumpeterian approaches, in terms of cycle duration, the timing of technological takeoff, technological opportunity, and the level of accumulated socio-economic capability or historical impact. These now almost stylised facts are also areas the chapter attempts to measure.

4.1.1. Chapter Outline

The organisation of the data upon which this analysis is based will first be presented. Then we set out to discuss the conceptual framework concerning the image of the shape of the technological growth curve, so as to develop an empirical model from which the data can be analysed. The cyclical units of the growth curve and the position of cyclical phases - depression, revival, prosperity and recession - will be identified.[4] As these are relative measures, we also need to bring in notions of a system's 'theoretical norm' and equilibrium. The periods of possible cyclical growth in technological development will then be identified, and emphasis will be on distinguishing sustained growth cycles from fluctuations. Although the conceptual framework will mainly be based on Kuznets (1930s, 1940), it is synthesised with the views of Schumpeter. Schumpeter was mainly a theorist, and his theoretical innovation cycles were not directly testable.[5] The chapter then presents the fitting of logistic growth

[4] Economic concepts of the phase or state of the innovation cycle is here used as a proxy measure, as the chapter builds upon Schumpeter's first approximation (prosperity and recession) and second approximation (depression and revival) which broadly refer to innovation investment and hence innovative activity (Schumpeter 1939, Chapter IV). It can be compared to Kuznets's primary trends in production and prices which reflect the life cycle of a given technical innovation (Kuznets 1930a).

[5] Kuznets and Schumpeter share some striking similarities and some equally striking differences. Whereas Schumpeter was a theorist and Kuznets an empirist they agreed that changes in technique are the decisive factor in growth. However, for empirical measurement purposes, Kuznets conceptual framework differed rather sharply from Schumpeter, although the two were not inconsistent with one another (Rostow 1990: 233-46). Although they both agreed on the existence of *S*-shaped growth paths, Schumpeter (for theoretical reasons) focused on the movements from 'prosperity and

curves to patent data. The use and transformation of the logistic biological growth model of the type used by Griliches (1957) is justified. There are numerous types of models which give rise to *S*-shaped paths, but they are quite different in the mechanism which actually produces this trajectory. We then examine whether technological development in different technological fields tends to follow *S*-shaped curves, and if so, the characteristics and properties of such curves are identified quantitatively. Each cycle will be characterised in relation to the parameters that guide its evolutionary path.

When identifying the diversity of technology dynamics, each cycle will be characterised in relation to the four phases of development (depression, revival, prosperity and recession), their timing as well as total cycle duration. The cycle's socio-economic capability or impact at different cyclical stages as well as the cycle's overall technology opportunity are also identified. After a cross technology comparison of the timing of technological takeoff and cycle saturation, the notion of clustering of innovations will also briefly be tested. (A much more involved examination of the possibility for innovation clusters follows in Chapter 5.) Some concluding remarks will follow in the last Section.

4.1.2. Data Selection

The quantitative analysis in this chapter is based on accumulated patent stocks of 399 US patent classes, as calculated in Chapter 2. For the purpose of this chapter each accumulated patent stock for each year 1920-90 is allocated to one of the 56 technological groups which collect together technologically related fields or sectors. It is at this 56 technological groups-level the analysis is conducted. In turn they belong to five broad technological spheres consisting of Chemical, Electrical/electronics, Mechanical and Transport fields as well as an 'other' group which is designated the Non-industrial group.

---

recession' to 'depression and revival', whereas Kuznets (for technical statistical reasons) focused on the movements from 'depression and revival' to 'prosperity and recession' (i.e. from trough to peak, see also section 4.3) (Kuznets 1940). Also, whereas Schumpeter begins with a theory of an economic dynamic system of circular flows into which he introduces a disruptive, creative, innovating entrepreneur, Kuznets begins with a set of observations (long time series) based on empirical evidence and statistical methods. Finally, whereas Kuznets asserted that *S*-shaped curves were randomly distributed, Schumpeter believed that they were clustered (Kuznets 1940).

4.2. THE *S*-SHAPED IMAGE OF THE TECHNOLOGICAL GROWTH CURVE[6]

4.2.1. The Conceptual Framework

The aim of this section is conceptual, dealing with how best to identify cyclical units (i.e. when a cycle starts and ends) and to identify the cyclical phases and saturation levels, in order to perform empirically the model fitting of patent stocks to *S*-shaped growth functions. This is an area where there is still no dominant approach.

In the definition of cyclical units, this chapter uses troughs and peaks as turning points (commonly used by national statistical offices). Between the turning points, indicating the growth cycle from trough to peak, the cycle passes through four cyclical phases of development, moving from depression to revival to prosperity and to recession. Instances of negative growth will be disregarded, and will instead be referred to as periods of crisis.

Hence, the analysis is based on Kuznets's conceptual approximation of a cycle using troughs and peaks as turning points. There is a dispute between Schumpeter and Kuznets concerning how to define the cyclical units of analysis. Schumpeter suggests that cyclical units should be defined from the beginning of prosperity (the point at which the time series begins to rise above the 'normal' level) to the end of revival (the point at which the time-series again reaches the new 'normal'). Kuznets stressed the empirical difficulty of developing statistical procedures that would correspond to Schumpeter's theoretical model: 'By refusing to accept peaks and troughs as guides in the determination of cycles he [Schumpeter] scorns the help provided by statistical characteristics of cycles in time series. One cannot escape the impression that Professor Schumpeter's theoretical model in its present state cannot be linked directly and clearly with statistically observed realities' (Kuznets 1940, p 116). '[N]o proper link is established between the theoretical model and statistical procedure' (Kuznets 1940, p.123). That is, we have technically developed statistical growth

---

[6] Throughout, the chapter refers to innovation 'cycles' and the 'cyclical' character of innovations with respect to *S*-shaped growth patterns (the appropriate application of such phrases are discussed in Solomou 1986, Kleinknecht 1990). The use of the term 'cycle' in this chapter is justified by the subsequent application of an evolutionary model endogenously producing *S*-shaped (cycle-like) growth patterns. Hence they are investigated as 'real' phenomena rather than purely statistical coincidence. However, the analysis does not to the same extent establish the 'cyclical' character of innovation cycles (i.e. their endogenously caused regular recurrence), so the chapter admittedly uses this concept in relation to a post-Schumpeterian theoretical belief on the behaviour of many economic variables.

functions which illustrate a path from trough to peak, as opposed to Schumpeter's growth functions illustrating evolution path from the beginning of prosperity to the end of revival. Schumpeter's and Kuznets's conceptual approximations of innovation cycles are illustrated in Figure 4.1.

The cyclical phases (i.e. depression, revival, prosperity, recession, crises) or stages of the cycle introduced by Schumpeter represent a relative measure. As Schumpeter argues, the different cyclical phases of development must be around something and, if pressed, that something is more or less related to an equilibrium concept (Schumpeter 1939, p. 44). For analysis and diagnosis purposes Schumpeter refers to equilibrium in a dynamic context (as opposed to the Walrasian system) in which any stage of a cycle can conveniently be defined in relation to its distance from a theoretical normal or 'theoretical norm' (Schumpeter 1939, p.23). Kuznets also acknowledges making business cycle theory an integral part of an economic system by coupling it to the achievements of equilibrium economics: 'Instead of a system we are likely to find ourselves with a chaotic description' (Kuznets 1930b, p.31), and he agreed with Schumpeter that the stable static equilibrium has to be substituted with a general recognition of the time element.

The growth cycle, wrapping around the system's theoretical norm, moves from one level of equilibrium to another at a qualitatively different level. If the troughs and peaks of every two successive growth cycles coincide in time, the technological development may have a pulsating character causing punctuated equilibrium (in the way argued by Yakovets 1994, p.400), although it is also argued here that this may not always be so if there is a long period of a crisis or a structural break when developing and adjusting to a new technological trend. As stated by Schumpeter, the equilibrium reached by recovery is not necessarily identical with that which would have been attained had the crisis not taken place.

As growth cycles between equilibrium points both for statistical technical purposes and theoretical purposes needs to be understood individually (c.f. each growth cycle being of different qualitatively kind - the nature of each cycle being unique), an offset - provided by a constant K - is inserted in the beginning of each cycle to avoid the previously accumulated patent stock entering the analysis. Hence, K reflects the size of the accumulated patent stock for each cycle's initial level, and is therefore of different value for each new cycle (see Figure 4.1, as well as section 4.3.2.).[7]

---

[7] It is relevant to mention that, as this chapter is based on a patent data classification scheme, we only measure technological activity in a certain technological area rather than a specific innovation. However, the type and nature of the technological activity varies over time.

Figure 4.1. Conceptual framework

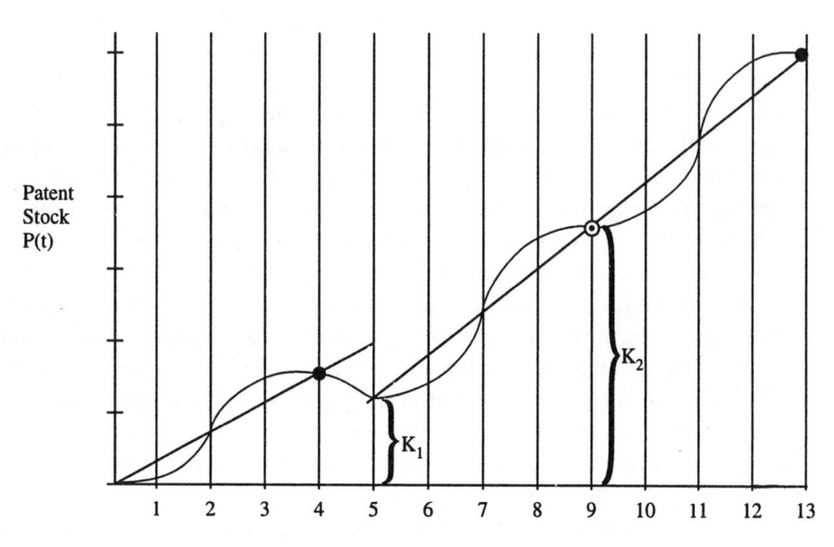

PHASES OF GROWTH

0–1; 5–6 and 9–10:	Depression
1–2; 6–7 and 10–11:	Revival
2–3; 7–8 and 11–12:	Prosperity
3–4; 8–9 and 12–13:	Recession
4–5:	Crises
2–7 and 7–11:	Illustrative examples of Schumpeter's innovation cycles
0–4, 5–9 and 9–13:	Illustrative examples of Kuznet's innovation growth cycles (the one adopted in this analysis)

Equilibrium (or level of saturation): ●
Punctuated equilibrium: ⊙
K: Offset

——— :The system's 'theoretical normal'

‿ or ⁀ : The innovation cycle

4.2.2. Identification of Periods of Potential Technological Growth Cycles

When identifying periods of potential technological growth cycles, it is important to distinguish between sustained cyclical growth periods and random fluctuations.

Using troughs and peaks as turning points, a growth cycle is considered to end or move into a crisis when two consecutive years display negative growth. The year before a sequence of two such negative growth observations is considered to be a peak, while the last observation of two consecutive years is considered to be a trough. By definition, a trough for a new growth cycle can at the earliest come two years after a peak of an earlier growth cycle, as we allow for one-year fluctuations within the growth cycles. The principle is taken from Franses (1996, p.85) and Watson (1994, p.24-46), who determine turning points in US industrial production data while investigating seasonality in business cycle turning points. Thus, in the context of this analysis, if a trough for a new growth cycle comes two years after the peak of an earlier cycle, we have defined this as a series of punctuated equilibria. However, if a trough for a new cycle comes more than two years after an earlier cycle, we have a series of negative growth rates reflecting a crisis.

Using this principle, the turning points in technological development of the accumulated patent stocks (calculated in Chapter 2, section 2.2) of 56 technological groups (presented in Chapter 2, section 2.4) have been identified and will subsequently be used for model fitting to test for the existence of *S*-shaped growth paths. Periods of sustained decline in the patent stock (i.e. periods of crisis, see Figure 4.1) are cut out of this analysis, and the remaining periods illustrate potential growth cycles. As much as 172 sustained growth periods were found across all the 56 technological groups and the structure of those are illustrated in Table 4.1. (Documentation of all the 172 observed sustained growth periods is in the Appendix (section 4.6, Tables 4.7 to 4.11) at the end of this chapter, which include all identified periods of sustained growth from trough to peak, across all technological groups.)

Of the 172 sustained growth periods, 6 technological groups of the 56 in total experienced only 1 sustained growth period, 14 groups experienced 2 sustained growth periods, 13 experienced 3, 17 experienced 4, 5 experienced 5, and finally, 1 technological group experienced 6 sustained growth periods between 1920 and 1990. Hence, the technological groups, identified in Table 4.1, show a notable difference in the number of growth periods.

Table 4.1. Structure of sustained growth periods between turning points

Numbers of technological groups by broad technological field	Number of 'technological groups' with sustained growth, between 1920 and 1990, for more than 40 / 50 / or 60 years in total, respectively	Total number of 'sustained growth periods from trough to peak' across all technologies / (sustained growth periods of at least five observation years)	Total number of 'sustained growth periods linked with punctuated equilibria' across all technologies. (assuming they can be fitted to *S*-shaped growth cycles)
Chemical: 13	12 / 10 / 7	31 / (25)	6
Electrical: 11	9 / 7 / 4	38 / (32)	6
Mechanical: 21	12 / 7 / 1	66 / (56)	15
Transport: 7	3 / 0 / 0	23 / (16)	4
Non-industrial: 4	1 / 0 / 0	14 / (11)	6
TOTAL: 56	37 / 24 / 12	172 (140)	37

Table 4.1 also indicates that in general we have longer periods of growth than periods of decline (which are left out), because (i) the total periods of sustained growth are longer than the total periods of sustained decline, and (ii) we often have one growth period after another, suggesting a series of punctuated equilibria between growth cycles. Of the 56 technological groups in total, 37 experienced sustained growth between 1920 and 1990 (a total of 70 years) for more than 40 years in total, and among those 37 groups, 24 experienced sustained growth for more than 50 years, and 12 even experienced sustained growth for more than 60 years. Hence, it is not surprising that we also find that as many as 37 of the 172 sustained growth periods can be linked with punctuated equilibrium (i.e. one growth cycle after another without any decline in accumulated patent stock) assuming they subsequently can be fitted to *S*-shaped growth cycles. Accordingly, (i) and (ii) above also indicate that the critique often raised, that technological growth curves used for picturing technological development only depict half of the technology's 'life', as they ignore the period of absolute technological decline (see e.g. Lindqvist 1994), is proven not to be the case when using patent data as a source of evidence.

In the further analysis only those sustained growth periods which have at least 5 observation years will be used (as this is the minimum time series number for fitting an *S*-curve to the data). Of these, Chemicals represents 25 potential growth curves, Electrical/electronics 32, Mechanical 56, Transport 16 and the Non-industrial technological group 11, which gives a total of 140 potential growth curves.

4.3. FITTING *S*-SHAPED GROWTH PATHS

4.3.1. Model Selection

Just as there is no dominant approach to the study of *S*-curves, there are no conclusive *S*-curve models associated with definite statistical tests. There are numerous types of models which give rise to *S*-shaped curves (see e.g. Mahajan and Peterson 1985), but they are quite different in the mechanism that actually produces this trajectory. Each development or diffusion model usually facilitates a theoretical explanation of the dynamics of the development or diffusion process in terms of certain general characteristics, as well as permits predictions concerning the continued evolution of the development or diffusion process. Several different kinds of such models have been applied in economics.

For the purpose of this study the biological internal-influence (i.e. endogenous) growth model, which has been applied by, e.g. Griliches (1957), is most applicable. The advantage of this type of internal-influence biological growth model is that it illustrates a growth process in which technology is the variable that grows over time and is not of a constant size which then diffuses, as in the internal-influence epidemic diffusion model (e.g. the one applied by Mansfield 1961). However, the disadvantage of the simple internally-influenced diffusion model is that it is of limited flexibility in terms of the two alternatives (i.e. symmetry or positively skew), as opposed to the flexible cumulative lognormal density models (e.g. used by Bain 1964).[8] That is, they are generally built around a logistic or a Gompertz growth curve.

As Kuznets (1930a, p.197), in his empirical research, finds that the simple logistic model, (as compared to the Gompertz model (e.g. used by Chow 1967)) best describes long-term movements of growing industries, and subsequently links these results to the technology life cycle (using cumulative stocks of US patent data) the logistic model is the most obvious to consider. The logistic model is also the most widely used, e.g. by Griliches (1957), who studied the output growth related to the diffusion of hybrid seed corn among

[8] The limitations of the flexible density model is however that, although it can deal with any skewness, it is basically a distribution function providing information concerning when you have the highest probability for growth or diffusion, but it does not provide any explanation concerning the mechanism behind the *S*-shape. It could be argued that any density model would fit any smooth movements in long time series, but as it does not provide any explanation concerning the mechanism of the movements, the results concerning the highest probability for growth or diffusion may be a pure illusion (a statistical coincidence). Also, in relation to this present study the density model is inappropriate, as we do not in all cases have the data from the initiation of the technological growth curve all the way to its long-run potential.

farmers in the US, and by Mansfield (1961) who studied the diffusion (or rate of imitation) of several industrial innovations among major firms. Metcalfe (1981) used the logistic curve in his extended model of innovation diffusion, which built partly upon the work of Kuznets's theory of industrial growth and retardation; and Andersen (E.S. Andersen 1994, Chapter 3) illustrated the way in which the logistic equation can provide us with a rough theoretical scheme which may help to organise the study of several of the issues raised by Schumpeter.

As the logistic function is symmetric around the inflection point the growth of the patent stock (or innovation growth) in the internal-influence growth model is proportional to the growth already received and to the distance from the ceiling.[9]

4.3.2. Model Fitting

Accordingly, the logistic internal-influence biological growth model is very useful when examining time series of patent data, and the parameters in the model are subject to simple and meaningful interpretation. The fundamental logistic internal-growth model as provided by Griliches (1957, p. 504) and applied to patent growth, is given here by:

(i) $$P(t) = \left[\frac{\overline{P}}{1 + \exp[-(a + bt)]} \right]$$

P(t) represents the accumulated patent stock at time t, $\overline{P}$ represents the ceiling (or long-run equilibrium) value, 'b' represents the rate of growth coefficient, and 'a' is a constant of integration which positions the curve on the time scale.

As Griliches (1957), Mansfield (1961), Bain (1964) etc. worked with diffusion expressed in percentage terms (and not real values), their model's initial stage, $P(t_{initial})$, is by definition close to zero, so the simple logistic model could be applied directly. The technological growth model will also subsequently be constructed along similar lines, by transforming the time-series of patent stocks for each growth curve into a percentage share of the long-term technological potential or ceiling.

However, when constructing the model we can gain a better impression of the diversity of technology dynamics (see section 4.4), derived from the cycle properties, if the cycle's origin, takeoff point, inflection point, saturation level, and phases of depression, revival, prosperity and recession are expressed in

[9] It is the last property which has made the logistic function useful in relation to Wolf's law of retardation, the importance of which was stressed by Kuznets.

terms of patent stocks and time. In this way it is also possible to see by how much the technological field has increased in absolute importance or impact during the growth cycle, and it is possible to estimate directly the duration of the growth cycle. This is done by adding an offset K to the growth function provided in equation (i) (K being a constant and different for each cycle), so that when calculating the accumulated patent stock in absolute terms, $P(t_{initial}) = K$ reflects an initial level of zero development (see also Figure 4.1). In this way, P(t) for any stage of development will always reflect a true value of the accumulated patent stock. Also, as the cycles do not start at year 0, a variable t_0 (t_0 being different for each cycle) has been inserted to represents the year of the first empirical observation of each growth curve, as listed in Table 4.1. In this way the estimated t value for each stage of development will represent a true year. Hence, the fundamental logistic biological growth model of patent growth over time is here rewritten as:

(ii) $$P(t) = \left[\frac{\overline{P}}{1 + \exp[-(a + b(t - t_0))]} \right] + K$$

where $\lim P(t) \approx (\overline{P} + K)$ for $t \to \infty$

P(t) represents the accumulated patent stock at time t. As the cycles does not start at year 0, t_0 (t_0 being different for each cycle) represents the year of the first empirical observation of each growth curve. $\overline{P}$ represents the cycle ceiling value, 'b' represents the rate of growth coefficient, 'a' is a constant of integration which positions the curve on the horizontal time scale. K is the constant of integration which positions the curve on the vertical patent stock scale. (K is the size of the accumulated patent stock at which the curve is initiated (i.e. ($t_{initial}$) or P(t) for $t \to -\infty$).)

Computational statistics based on non-linear least squares regression are then used to fit logistic curves to the sustained growth periods identified in section 4.2.2. In this process the computer simultaneously optimises or fits the parameter values of the origin or K, the cycle ceiling $\overline{P}$ (from which we can calculate $P(t_{ceiling}) = (\overline{P} + K)$), the integration coefficient on the time scale 'a', and the growth coefficient 'b'.

4.3.3. Model Fitting Results: The Existence of *S*-shaped Growth Paths.

The results of the parameter estimates can now be used to identify the characteristics, or cycle properties, of the *S*-shaped growth paths (identified in section 4.2.2) successfully fitted to the logistic biological growth model; subject to a few constraints. *First*, the goodness of fit, R^2, has to be greater than 0.94. *Second*, it might be expected that the technological growth cycle will be interrupted or be the subject of structural breaks due to random historical events before its estimated long-run performance level or full potential is reached. Thus, another constraint is that an empirical break from the growth cycle model (i.e. the time of the last empirical observation of the cycle) has to be at least after 60% of the long-run estimated technological cycle potential in terms of accumulated patent stock, i.e. $(P(t)-K)/\overline{P}$ has to be at least 0.6 (60%). This includes even those technological groups whose cycles break empirically in 1990 owing to data constraints. It is necessary to go beyond the point of inflection of 50% in order to make sure that we actually have an S-curve and not exponential growth. A *third* constraint is that the accumulated patent stock at the first empirical observation of the cycle, i.e. t_0, has to be below 10% of the patent stock at the cycle ceiling (i.e. full estimated growth cycle). This means that we allow the technology to 'die' relatively young (already after 60% of its full potential) but, as in biology, it cannot by definition be 'born' old in terms of empirical observations. However, as there is no reason to believe that these technology growth cycles first entering in 1920 due to constraints in the data source were actually born in 1920, we allow them into the model as long as 1920 is still in the first phase of the growth cycle, i.e. below 25% of the full cycle (i.e. $(P(t)-K)/\overline{P}$ can be up to 0.25 (25%) in 1920), which also represents the initial growth phase before takeoff and revival.

Among the growth paths identified in section 4.2.2 that were eligible for estimation, the broad Chemical group went down by 4 (from 25 to 21), Electrical/electronics down by 9 (from 32 to 23), Mechanical down by 11 (from 56 to 45), Transport down by 8 (from 16 to 8), and the Non-industrial technological group went down by 2 (from 11 to 9). Hence 106 technological growth curves out of 140 survived the constraints, and could be fitted to the selected *S*-curve model. The results of the parameter estimates are listed in Tables 4.12 to 4.16 in the Appendix, (Section 4.6).

With respect to the first and third constraint, the results in the Appendix also show that the goodness of fit is usually at the 0.99 or 0.98 level (applying to 86 out of the 106 *S*-curves), and that the accumulated patent stock at the time of the first empirical observation is below 5% of the full technological potential for most of the growth cycles (applying to 65 out of the 106 *S*-curves).

Finally, as mentioned in constraint two, it might be expected that the technological growth cycle will be interrupted or be the subject of structural breaks due to random historical events before its estimated long-run performance level or full potential is reached. It is interesting here to compare the estimated accumulated patent stock at the time of the last empirical observation of the cycle (identified in the Appendix, (Section 4.6), Tables 4.7 to 4.11) with the estimated cycle potential (or cycle ceiling) to see to what extent the cycle reaches its long-run potential (identified in Tables 4.2 to 4.6). (The calculation of the percentages share of technological development in relation to the potential ceiling of the growth curve is represented in Section 4.4.) Evidence from such analysis shows that almost all cycles (i.e. 72 out of 106 in total) survive more than 90% of their long-run potential or ceiling level without any external shocks to break the cycle, and that it is very rare (i.e. only 10 out of 106) that they get interrupted before 80% of their estimated ceiling value. Actually, as many as 44 out of the 106 growth cycles represent a full cycle (i.e. more than 95% of their long-run potential or ceiling level) without any breaks. The estimated trajectories of those 44 cycles are plotted in Graphs 4.1 - 4.44 during their period of operation (see period used for model fitting, Tables 4.12 - 4.16). Here it ought to be mentioned that many of the cycles ending empirically in 1990 are most likely because of data constraints (as the compiled data used for this study range from 1890 to1990) than anything else. So although external factors and 'history' have a large influence on change and growth, it is still important to understand the mechanism and the internal system dynamics.

These factors indicate that we have good model-fitting results, which in turn justifies the use in evolutionary economics of the *S*-shaped image or metaphor concerning the way technology develops.

Graph 4.1. Chemicals (3a): Inorganic chemicals: [year; patent stock]

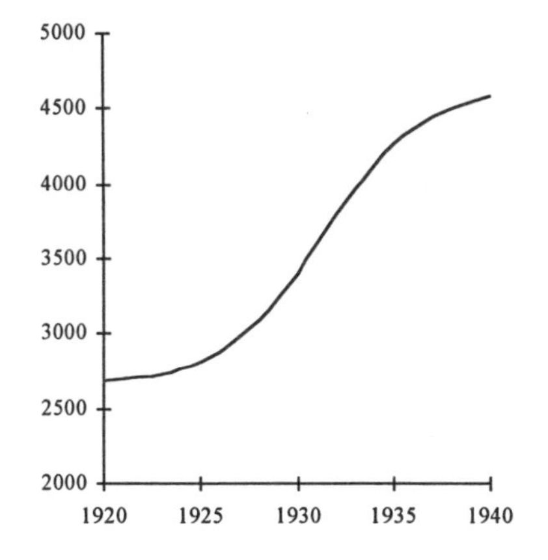

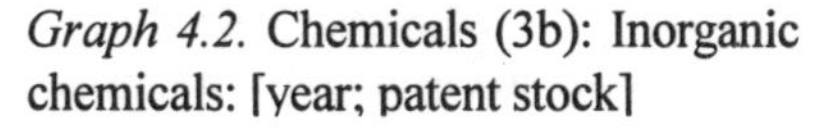
Graph 4.2. Chemicals (3b): Inorganic chemicals: [year; patent stock]

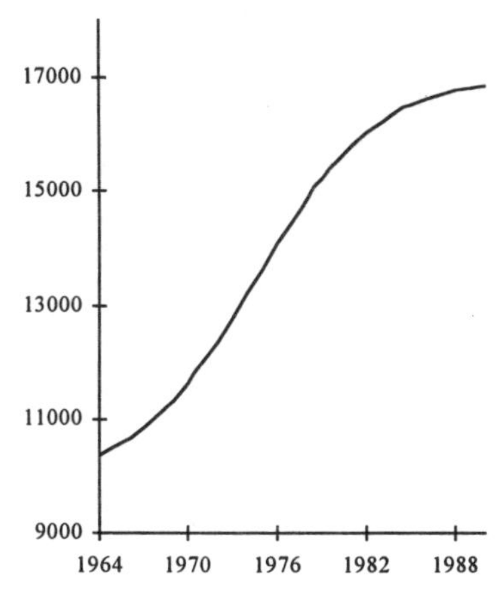

Graph 4.3. Chemicals (4a): Agricultural chemicals: [year; patent stock]

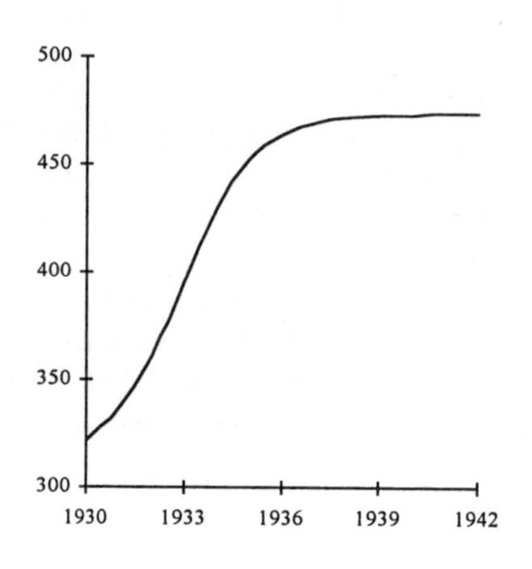

Graph 4.4. Chemicals (7c): Cleaning agents and other compositions: [year; patent stock]

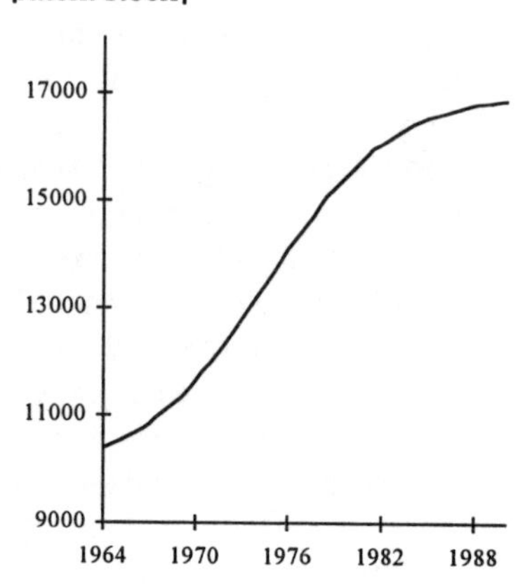

Graph 4.5. Chemicals (10b): Bleaching and dyeing: [year; patent stock]

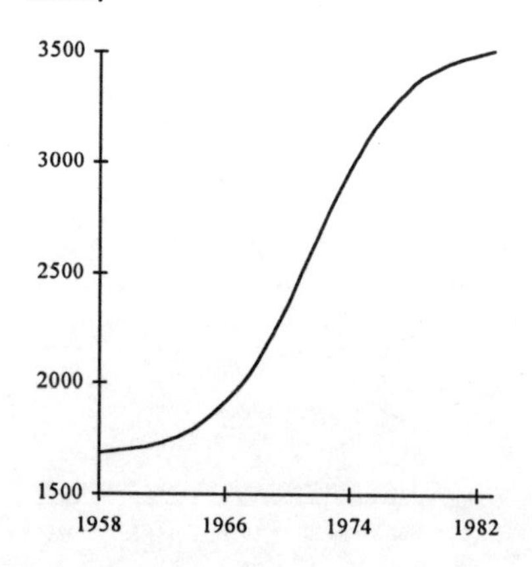

Graph 4.6. Chemicals (51a): Coal and petroleum products: [year; patent stock]

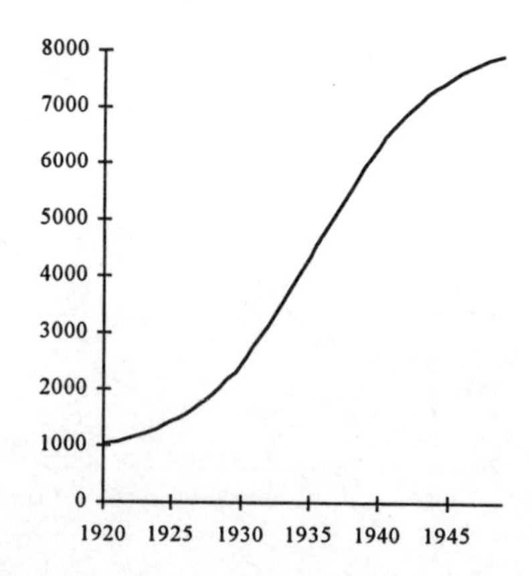

Graph 4.7. Chemicals (51b): Coal and petroleum products: [year; patent stock]

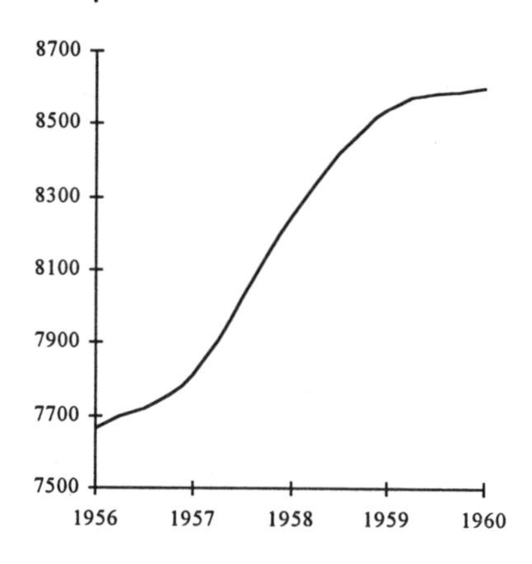

Graph 4.8. Chemicals (51c): Coal and petroleum products: [year; patent stock]

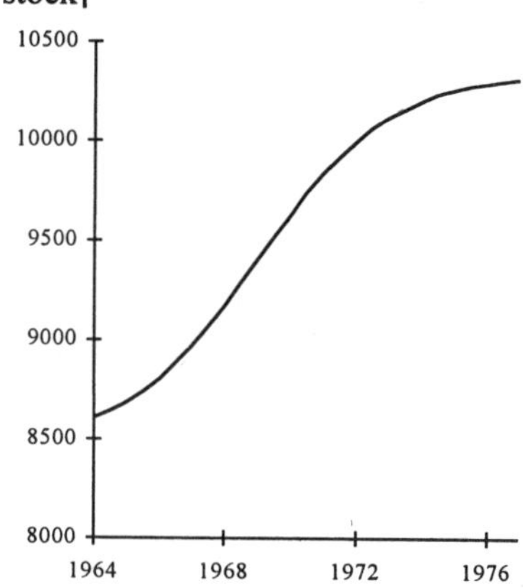

Graph 4.9. Electrical/electronics (30b): Mechanical calculators and typewriters: [year; patent stock]

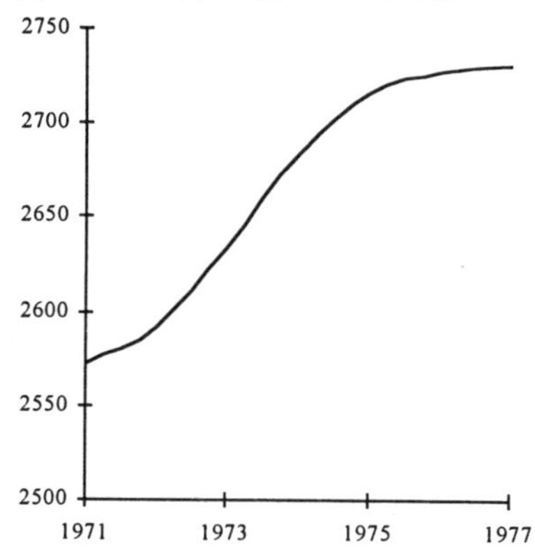

Graph 4.10. Electrical/electronics (33): Telecommunication: [year; patent stock]

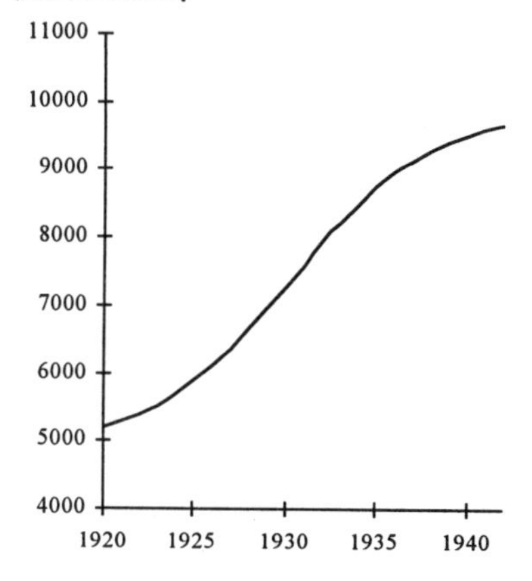

Graph 4.11. Electrical/electronics (37b): Illumination devices: [year; patent stock]

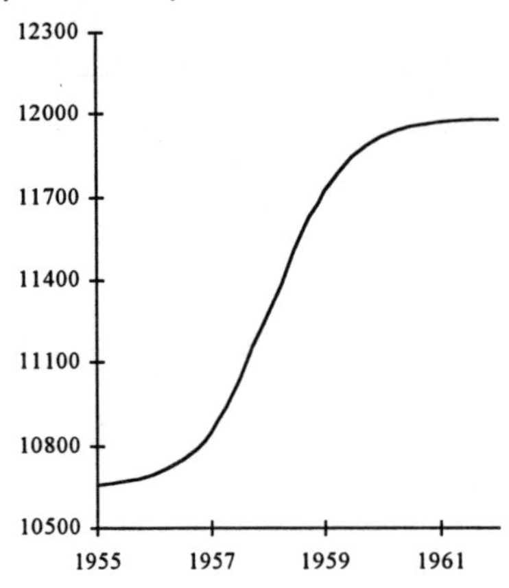

Graph 4.12. Electrical/electronics (38b): Electrical devices and systems: [year; patent stock]

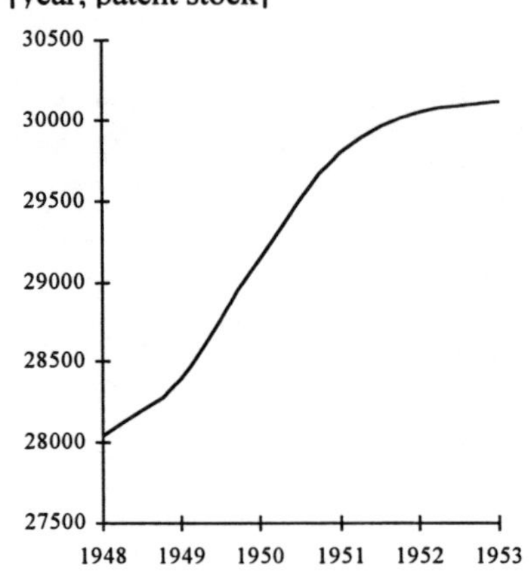

Graph 4.13. Electrical/electronics (39a): Other general electric equipment: [year; patent stock]

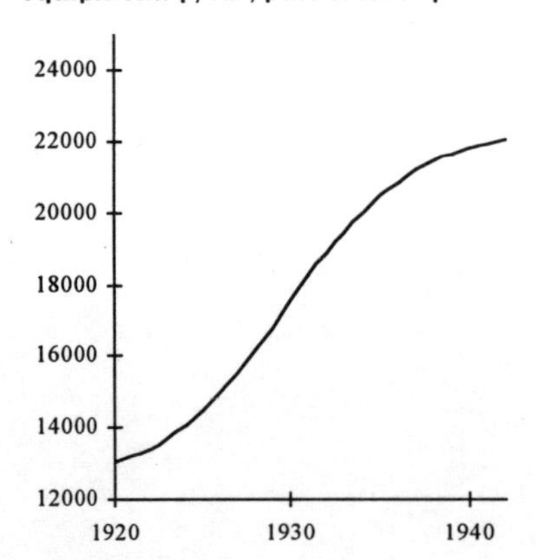

Graph 4.14. Electrical/electronics (39b): Other general electrical equipment: [year; patent stock]

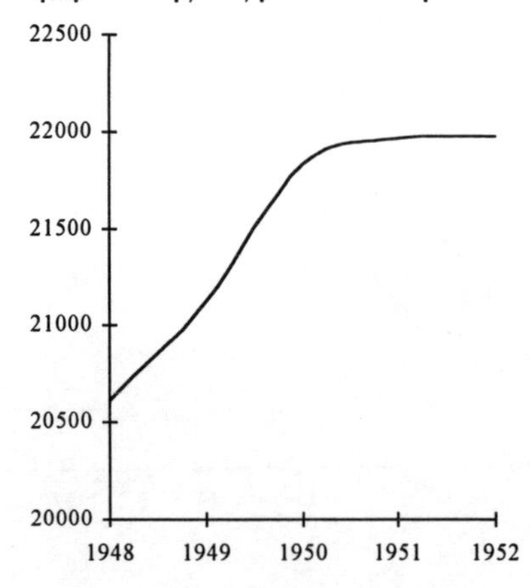

Graph 4.15. Electrical/electronics (39c): Other general electric equipment: [year; patent stock]

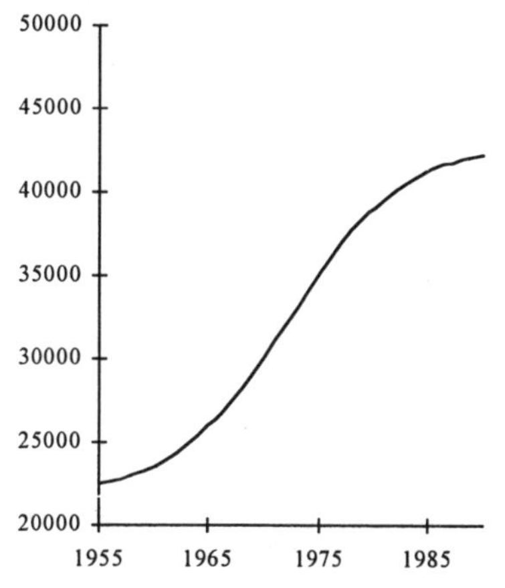

Graph 4.16. Electrical/electronics (40): Semiconductors: [year; patent stock]

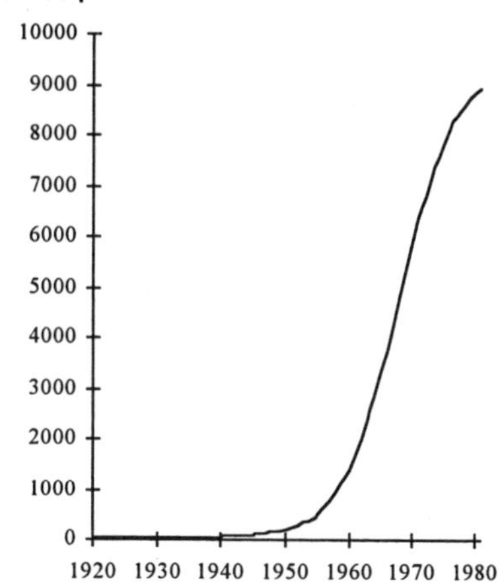

Graph 4.17. Electrical/electronics (52b): Photographic equipment: [year; patent stock]

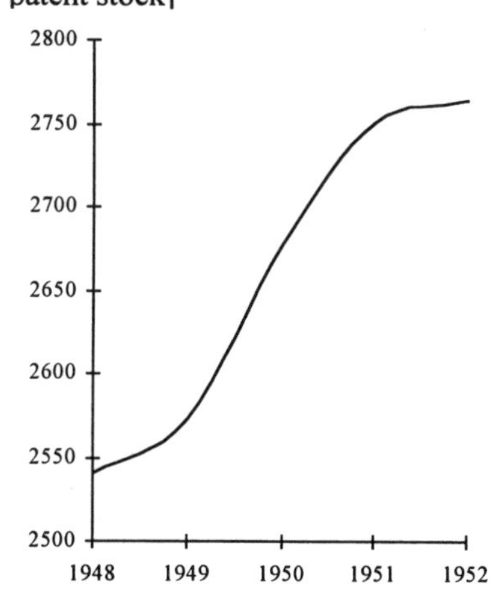

Graph 4.18. Mechanical (13b): Metallurgial processes: [year; patent stock]

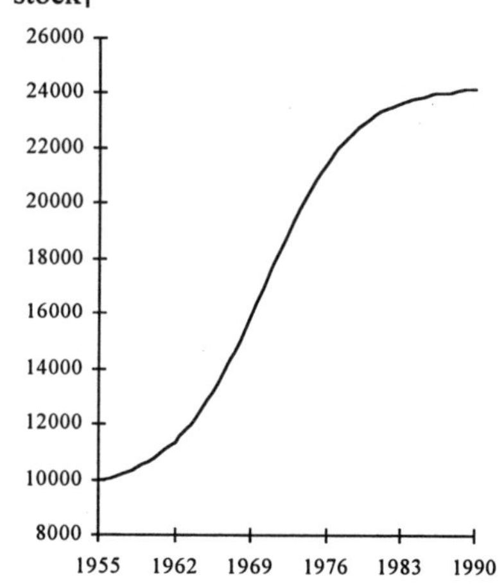

Graph 4.19. Mechanical (15a): Food, drink and tobacco equipment: [year; patent stock]

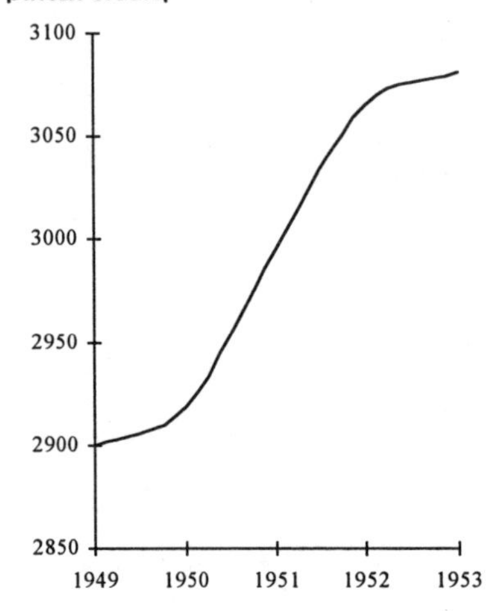

Graph 4.20. Mechanical (15c): Food, drink and tobacco equipment: [year; patent stock]

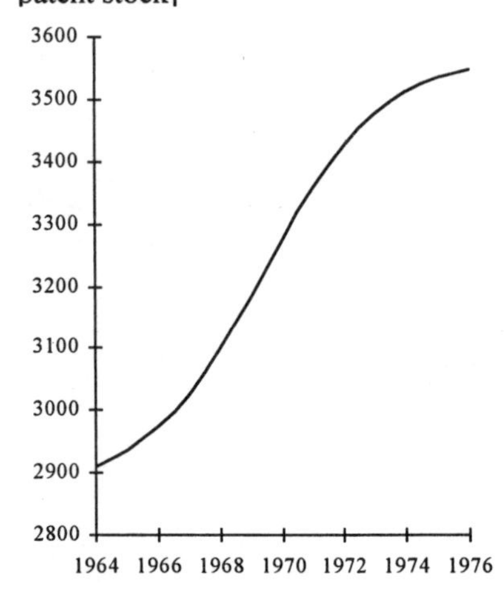

Graph 4.21. Mechanical (16b): Chemical and allied equipment: [year; patent stock]

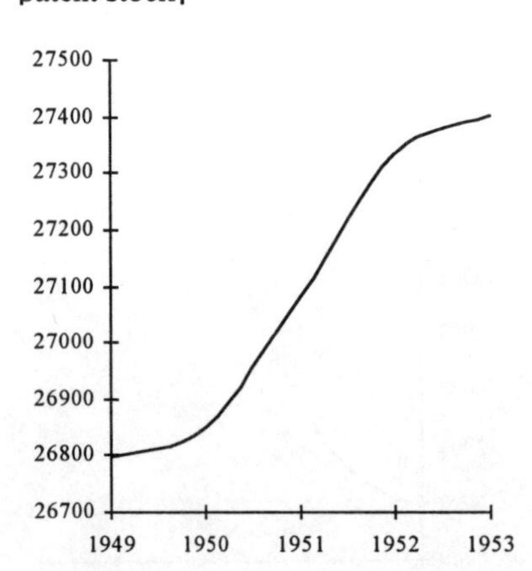

Graph 4.22. Mechanical (16c): Chemical and allied equipment: [year; patent stock]

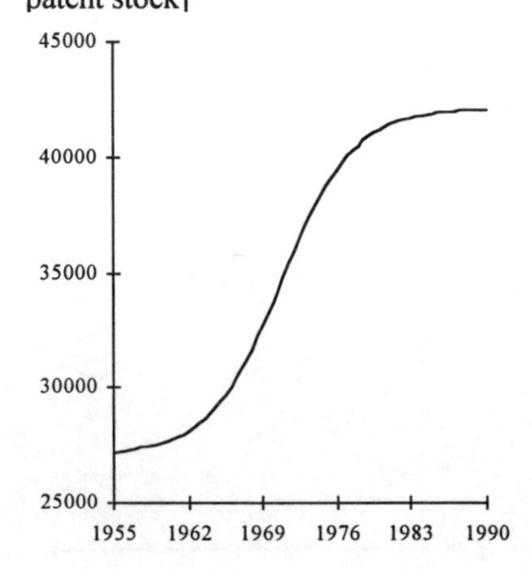

Graph 4.23. Mechanical (17b): Metal working equipment: [year; patent stock]

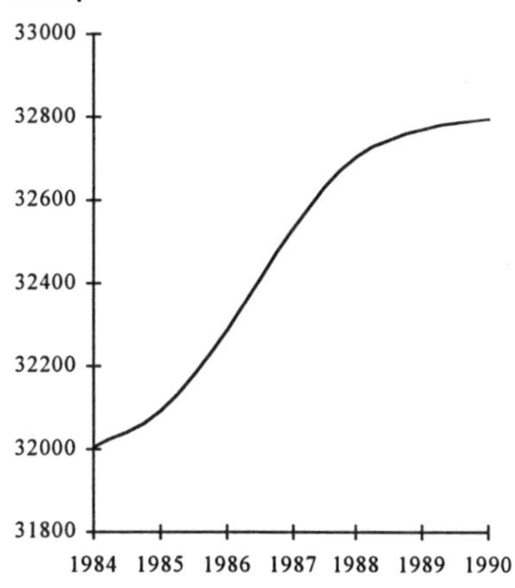

Graph 4.24. Mechanical (18a): Paper making apparatus: [year; patent stock]

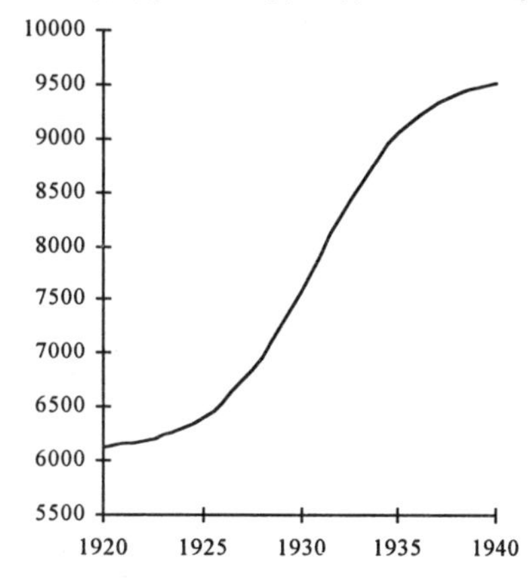

Graph 4.25. Mechanical (19a): Building material processing equipment: [year; patent stock]

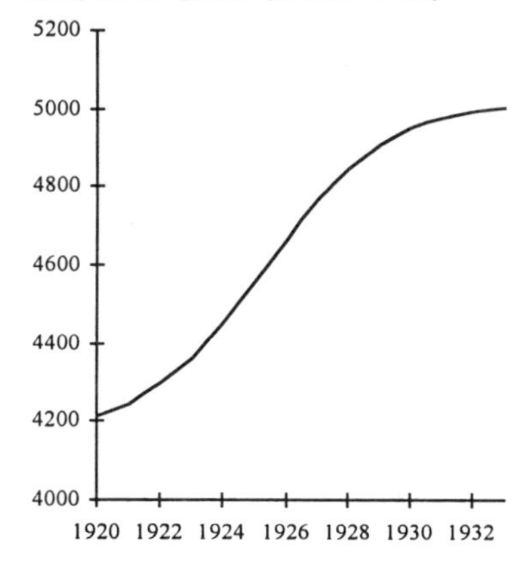

Graph 4.26. Mechanical (19b): Building material processing equipment: [year; patent stock]

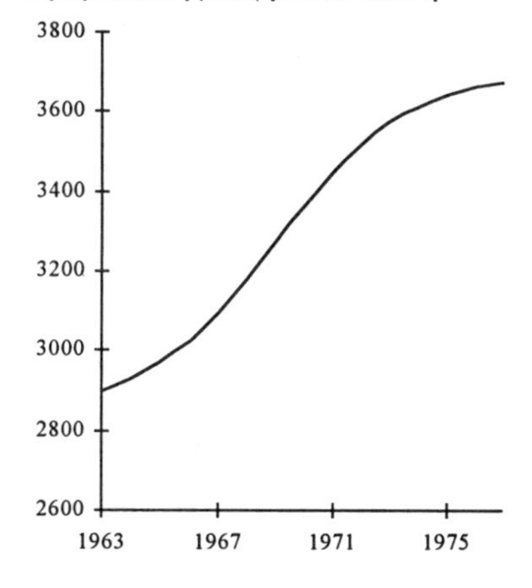

Graph 4.27. Mechanical (21a): Agricultural equipment: [year; patent stock]

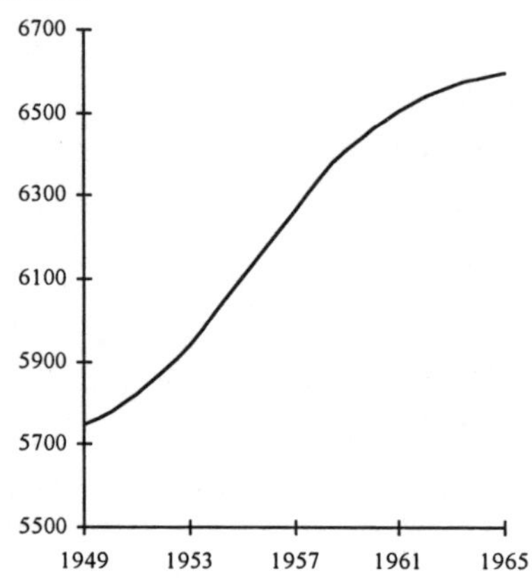

Graph 4.28. Mechanical (21b): Agricultural equipment: [year; patent stock]

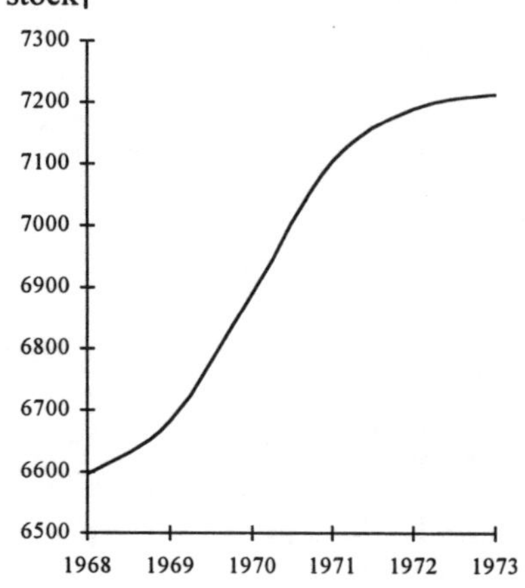

Graph 4.29. Mechanical (23a): Mining equipment: [year; patent stock]

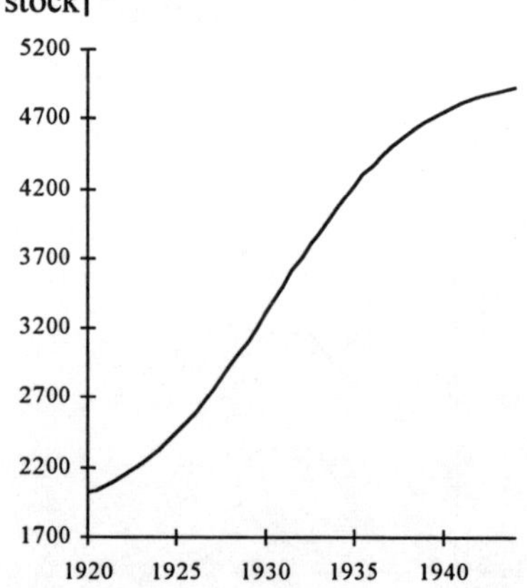

Graph 4.30. Mechanical (23b): Mining equipment: [year; patent stock]

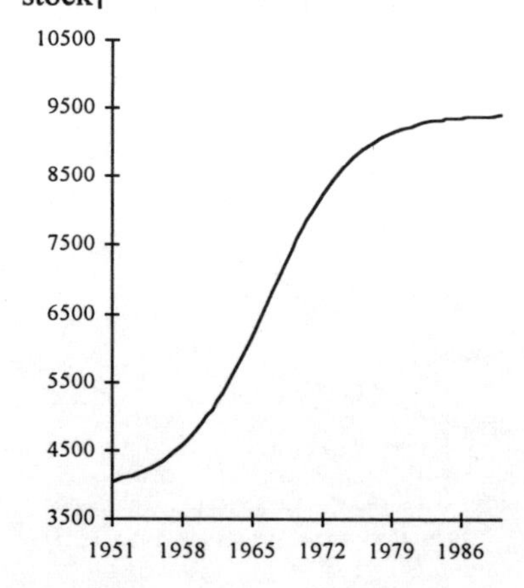

Graph 4.31. Mechanical (24a): Electrical lamp manufacturing: [year; patent stock]

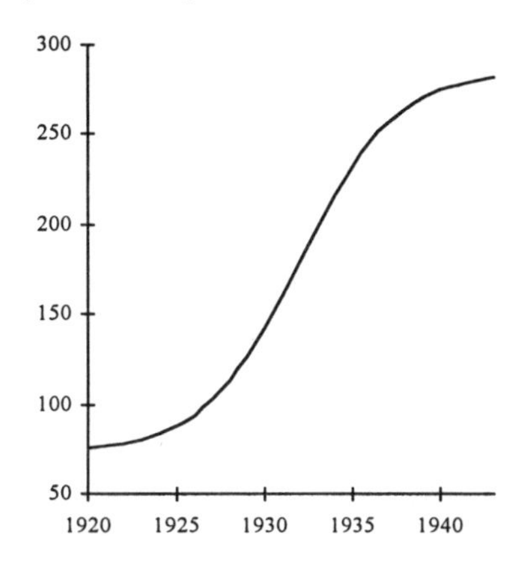

Graph 4.32. Mechanical (25): Textile and clothing machinery: [year; patent stock]

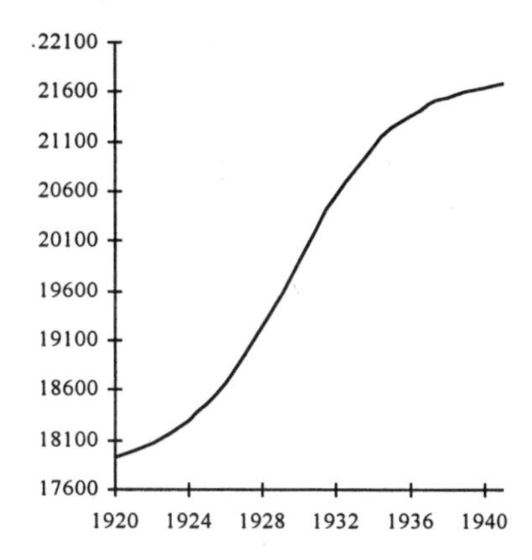

Graph 4.33. Mechanical (26a): Printing and publishing machinery: [year; patent stock]

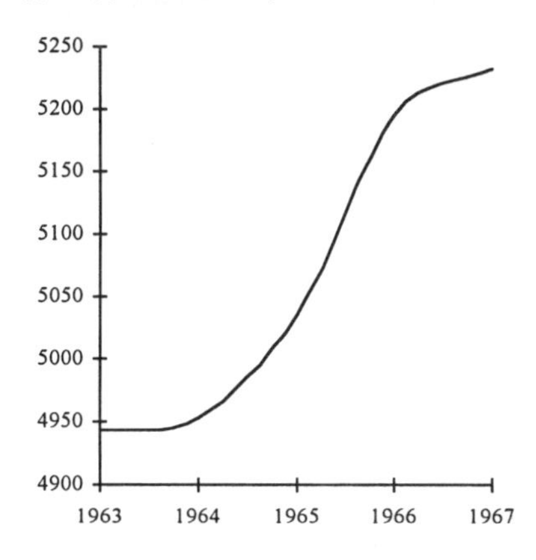

Graph 4.34. Mechanical (27a): Woodworking and machinery: [year; patent stock]

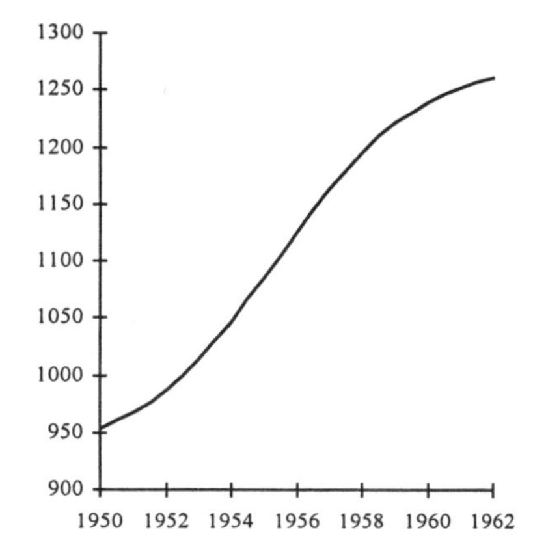

Graph 4.35. Mechanical (28b): Other specialised machinery: [year; patent stock]

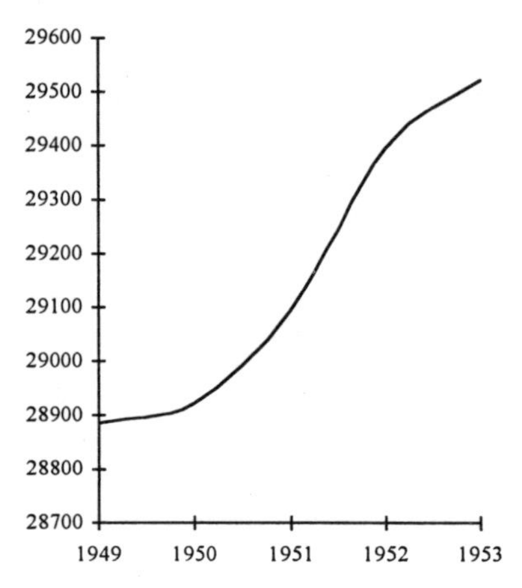

Graph 4.36. Mechanical (29b): Other general industrial machinery: [year; patent stock]

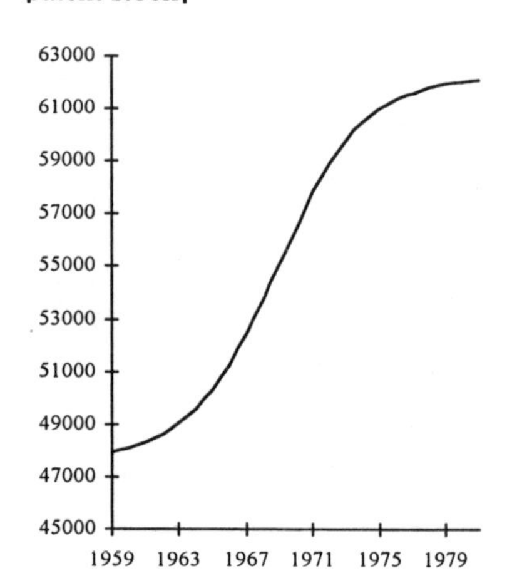

Graph 4.37. Mechanical (31): Power plants: [year; patent stock]

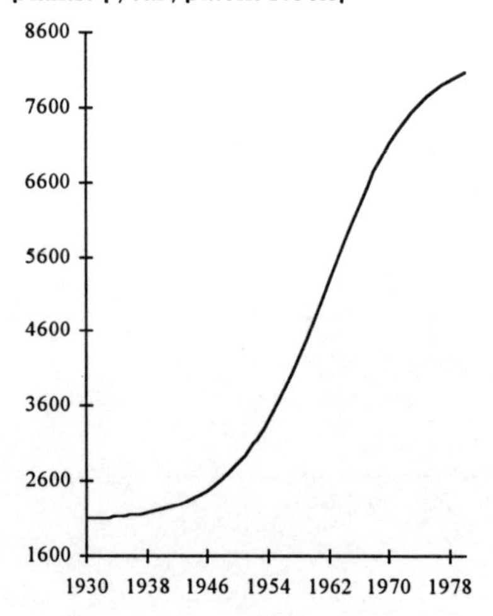

Graph 4.38. Mechanical (50a): Non-metallic mineral products: [year; patent stock]

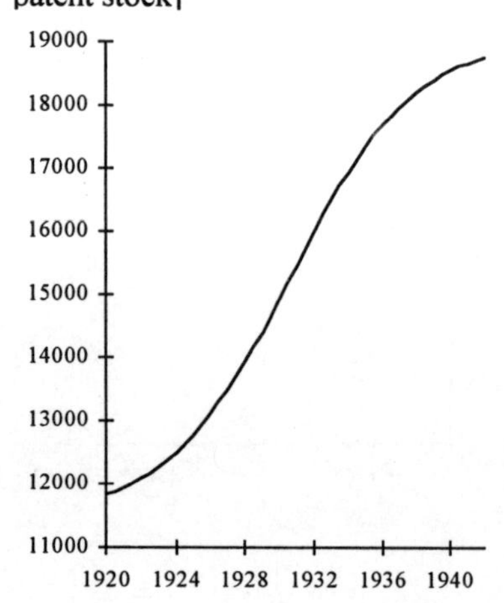

Graph 4.39. Transport (45a): Ships and marine propulsion: [year; patent stock]

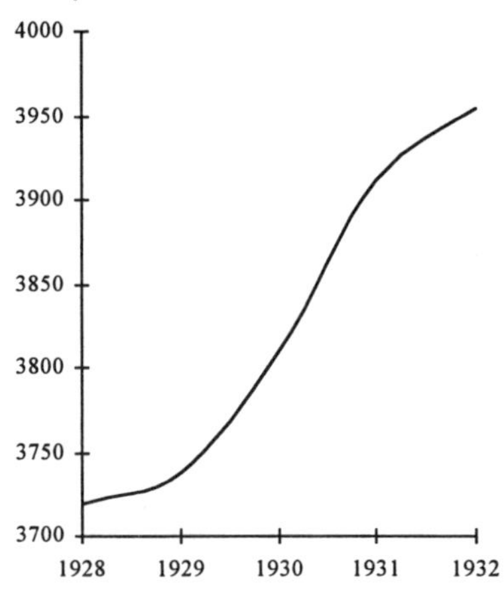

Graph 4.40. Transport (49a): Rubber and plastic products: [year; patent stock]

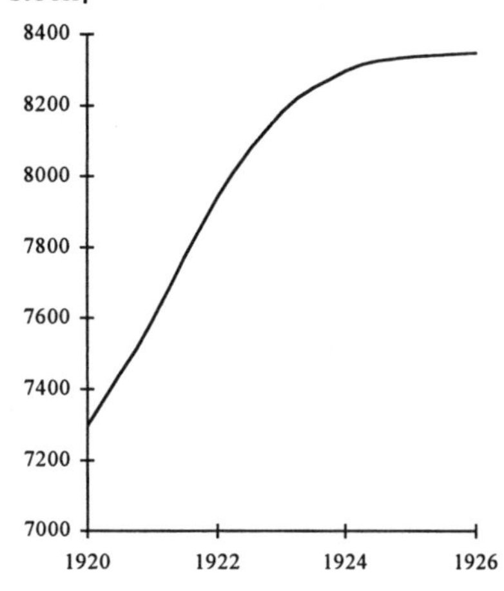

Graph 4.41. Non-industrial (32a): Nuclear reactors: [year; patent stock]

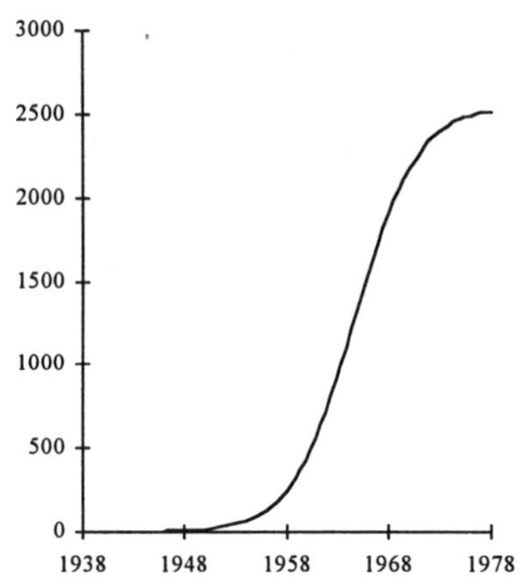

Graph 4.42. Non-industrial (32b): Nuclear reactors: [year; patent stock]

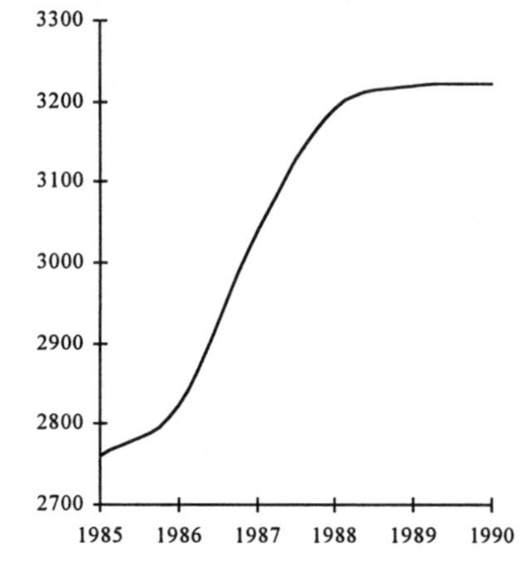

Graph 4.43. Non-industrial (56a): Other manufacturing and non-industrial: [year; patent stock]

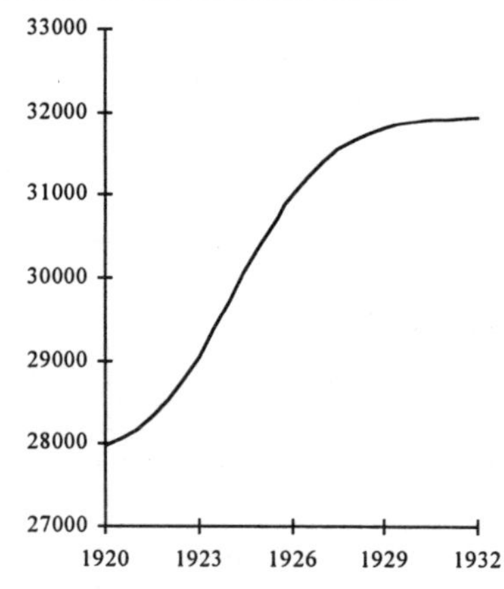

Graph 4.44. Non-industrial (56b): Other manufacturing and non-industrial: [year; patent stock]

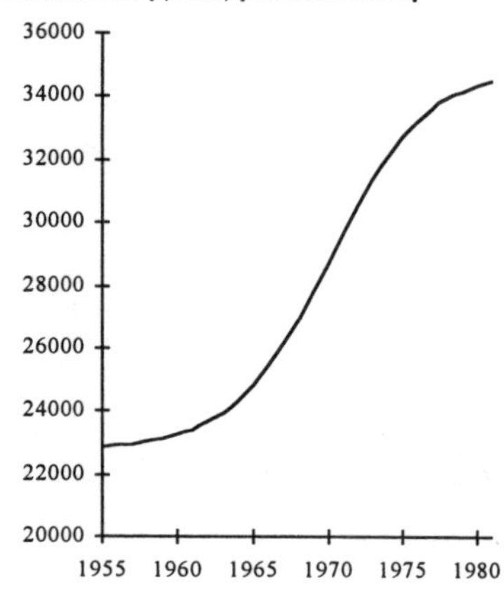

4.4. DIVERSITY OF TECHNOLOGY DYNAMICS

4.4.1. Fitting Cycle Properties

When fitting cycle properties (see '*Cycle properties*' below) concerning the diversity of technology dynamics (including timing and socio-economic capability or impact at the year of takeoff and for the four phases of the growth curves (depression, revival, prosperity and depression), as well as the cycle duration and technological opportunity of the cycle), the time and level of accumulated patent stock for any stage on the cycle first need to be identified. For this purpose, the growth model is transformed so as to calculate the share of the technology (i.e. patent stock) which has been developed in relation to its long-run level of technological potential or equilibrium. Hence, the technological relative position at a given time will always be in terms of the share of its long run potential.[10]

[10] Although Griliches (1957), Mansfield (1961), Bain (1964) etc. worked with diffusion or growth expressed in relation to a given total, in which the total represents 100%, their ceiling value was not necessarily at the 100% level. For example, Griliches (1957) had some land which would never get planted with hybrid seed corn so the ceiling level would be below 100%. Similarly, in innovation diffusion studies, some firms may never

By first subtracting K from P(t) (in order to scale the model so that $P(t_{initial}) \approx$ 0% of development), and then dividing both sides of the logistic model (see equation ii) by $\overline{P}$ (in order to consider the stage of the growth cycle in relation to the long-run level of its potential), we have the form:

$$\text{(ii)} \qquad \frac{(P(t) - K)}{\overline{P}} = \frac{1}{1 + \exp[-(a + b\,(t - t_0))]}$$

$$\text{where } \lim \left[\frac{(P(t) - K)}{\overline{P}}\right] \approx 1 \text{ (or 100\%) for } t \to \infty$$

Denoting 'x' as the stage, $(P(t)\text{-}K)/\overline{P}$, which the technology has developed out of its long run potential , the accumulated patent stock, P(t), for any given time, t, can then be calculated as $P(t) = (x)\,\overline{P} + K$. By rewriting equation (iii) and inserting P(t), the time t for any stage of development can be calculated by:

$$\text{(iv)} \qquad t = \frac{-\left(LN\left[\dfrac{\overline{P}}{(x)\overline{P}} - 1\right]\right) - a}{b} + t_0$$

E.g. by inserting P(t) for the point of inflection, x = 0.5, we get the time, t, of cycle inflection to be $-a/b + t_0$.

Cycle properties

As the logistic model is symmetrical around the inflection point, the model can now be directly associated with the following properties, concerning different cyclical points (initiation, takeoff and ceiling) and concerning the four phases of the growth cycle (depression, revival, prosperity and recession), which play an important part in identifying the diversity of technology dynamics:

- Initial level of technology growth curve, if $(P(t)\text{-}K)/\overline{P} \approx 0$ (0%)
- Depression, if $(P(t)\text{-}K)/\overline{P} < 0.25$ (25%) of the technological growth curve
- Takeoff of technology, if $(P(t)\text{-}K)/\overline{P} = 0.25$ (25%)
- Revival, if 0.25 (25%) $\leq (P(t)\text{-}K)/\overline{P} < 0.5$ (50%) of the technological growth curve
- Inflection point, if $(P(t)\text{-}K)/\overline{P} = 0.5$ (50%)

---

adopt a new technology so the ceiling will be below 100%. However, in the current model concerning the growth of technology, the ceiling level is by definition 100% representing the technology's full potential.

- Prosperity, if 0.5 (50%) $\leq$ (P(t)-K)/ $\bar{P}$ < 0.75(75%) of the technological growth curve
- Recession, if 0.75 (75%) $\leq$ (P(t)-K)/ $\bar{P}$ <1 (100%) of the technological growth curve
- Potential ceiling of technology growth curve, if (P(t)-K)/ $\bar{P}$ $\approx$ 1 (100%)

Besides the degree of synchronisation in time in terms of different cyclical points and phases of development, the diversity of technology dynamics is also characterised in relation to three of the most consulted properties relating to the structure of technological cycles: (i) Cycle *duration,* (ii) technological *opportunity* of cycle, and (iii) technological *socio-economic capability or impact* of the cycle. In this context:

- The potential duration of the cycles are calculated over the period in which the accumulated patent stock moves from 5% to 95% of the ceiling level, as we will only approach 0% when $t \to -\infty$ and 100% when $t \to \infty$.
- Overall potential technological opportunity of a cycle is expressed as the percentage growth in patent stock across the estimated cycle duration: $[P(t_{95\%})-P(t_{5\%})]*100\% / P(t_{5\%})$.
- Accumulated socio-economic capability or impact of a cycle at a given phase or time is expressed as the accumulated patent stock, P(t), at that time.

Chapter 2 (section 2.2) discusses the use of patent data as proxy measures for technological opportunity and socio-economic technological capability or impact.

4.4.2. Cycle Property Results

The results of the timing and socio-economic capability or impact at the year of takeoff and for the four phases of the growth curves, as well as the cycle duration and technological opportunity of cycle are displayed in Tables 4.2-4.6.

The results show absolutely no evidence of a fixed cycle duration, as the length of each cycle spans from just a few years up to several decades, and this is true for all broad groups: Chemicals, Electrical/electronics, Mechanical, transport and non-industrial. Sorting the 106 identified cycles in Tables 4.2 to 4.6 by cycle duration, we have: 34 cycles < 10 years; 10 years ≥ 27 cycles < 20 years; 20 years ≥ 25 cycles < 30 years; 30 years ≥ 10 cycles < 40 years, 40 years ≥ 4 cycles 50 years; and finally, 6 cycles with a cycle duration ≥ 50 years. Hence, we see that, the longer the cycles the more rare they become.

Table 4.2. Cycle properties: Chemicals

	Depression *		Revival		Prosperity		Recession			
	Socio-economic capability at:		Socio-economic capability at:		Socio-economic capability at:		Socio-economic capability at:		Socio-economic capability at:	Cycle Duration in
Code** / Technological group	P(t5%)*	Year	P(t25%)	Takeoff year	P(t50%)	Year	P(t75%)	Year	P(t95%)	Year	years	Cycle opportunity *** (Growth rate)

Code** / Technological group	P(t5%)*	Year	P(t25%)	Takeoff year	P(t50%)	Year	P(t75%)	Year	P(t95%)	Year	Cycle Duration in years	Cycle opportunity *** (Growth rate)
	*Depression **		*Revival*		*Prosperity*		*Recession*					
	Socio-economic capability at:		Socio-economic capability at:		Socio-economic capability at:		Socio-economic capability at:		Socio-economic capability at:			
2 Distillation processes	94.24	1900	456.01	1931	908.22	1949	1360.43	1967	1722.20	1998	98	1727.46 %
3a Inorganic chemicals	2760.71	1924	3157.18	1928	3652.76	1931	4148.34	1934	4544.81	1939	15	64.62 %
3b Inorganic chemicals	3900.09	1954	5062.29	1963	6515.05	1968	7967.81	1973	9130.02	1982	28	134.10 %
4a Agricultural chemicals	318.30	1930	350.96	1932	391.78	1933	432.61	1934	465.27	1936	6	46.17 %
4b Agricultural chemicals	619.36	1954	1511.31	1968	2626.26	1977	3741.21	1986	4633.17	2000	46	648.06 %
5a Chemical processes	4476.57	1921	5901.40	1928	7682.43	1932	9463.47	1937	10888.29	1944	23	143.23 %
5b Chemical processes	11829.40	1956	17701.50	1967	25041.70	1973	32381.90	1980	38254.07	1991	35	223.38 %
6 Photographic chemistry	1115.90	1947	4038.83	1966	7692.50	1977	11346.20	1989	14269.10	2008	61	1178.71 %
7a Cleaning agents and other compositions	3497.91	1922	4873.04	1931	6591.94	1937	8310.84	1943	9685.96	1953	31	176.91 %
7b Cleaning agents and other compositions	8980.24	1954	9301.22	1958	9702.44	1960	10103.70	1963	10424.64	1966	12	16.08 %
7c Cleaning agents and other compositions	10260.40	1963	11684.10	1970	13463.80	1975	15243.50	1979	16667.24	1986	23	62.44 %

Table 4.2. (cont.)

8a	Disinfecting and preserving	40.25	1915	54.72	1926	72.82	1933	90.91	1939	105.38	1950	35	161.81 %
8b	Disinfecting and preserving	122.25	1955	212.85	1968	326.10	1976	439.34	1984	529.94	1997	42	333.49 %
9	Synthetic resins and fibres	2727.71	1939	11488.70	1959	22439.90	1971	33391.10	1983	42152.03	2003	64	1445.33 %
10a	Bleaching and dyeing	725.26	1924	1095.25	1930	1557.74	1934	2020.22	1938	2390.21	1944	20	229.57 %
10b	Bleaching and dyeing	1758.39	1963	2135.39	1968	2606.66	1972	3077.92	1975	3454.93	1980	17	96.48 %
51a	Coal and petroleum products	1186.40	1923	2668.94	1931	4522.10	1935	6375.26	1940	7857.80	1948	25	562.32 %
51b	Coal and petroleum products	7686.60	1956	7881.22	1957	8124.49	1958	8367.77	1958	8562.39	1959	3	11.39 %
51c	Coal and petroleum products	8560.36	1963	8937.13	1967	9408.10	1969	9879.07	1971	10255.84	1975	12	19.81 %
55a	Explosive compositions and charges	412.25	1929	426.90	1930	445.22	1931	463.54	1932	478.20	1934	5	16.00 %
55b	Explosive compositions and charges	415.02	1957	635.23	1964	910.50	1968	1185.76	1973	1405.97	1980	23	238.77 %

*Phases of growth are expressed as % share of cycle developed in comparison to the cycle ceiling: $(P(t)-K)/\overline{P}$ (see section 4.4.1.). The time (year) and accumulated patent stock (or socio-economic capability or importance) for any such phase (i.e. for $(P(t)-K)/\overline{P}$ = 0.05 (5%), 0.25 (25%), 0.50 (50%), 0.75 (0.75%) and 0.95 (0.95%)) are calculated. ** a, b, and c, after the technological group code number, denote different cycles of same technological group across different periods in time (see Year columns). ***Cycle opportunity is expressed as the percentages growth in patent stock across the estimated cycle duration: $[P(t_{95\%})-P(t_{5\%})]*100\% / P(t_{5\%})$.

Table 4.3. Cycle properties: Electrical / electronics

Code** / Technological group	Socio-economic capability at: P(t5%)*	Year	Socio-economic capability at: P(t25%)	Takeoff year	Socio-economic capability at: P(t50%)	Year	Socio-economic capability at: P(t75%)	Year	Socio-economic capability at: P(t95%)	Year	Cycle duration in years	Cycle opportunity***: (Growth rate)
	Depression * (P(t5%) – P(t50%))											
			Revival (P(t25%) – P(t75%))									
					Prosperity (P(t50%) – P(t95%))							
							Recession (P(t75%) – Year at P(t95%))					
30a Mechanical calculators and typewriters	4711.09	1921	4773.59	1923	4851.70	1924	4929.82	1925	4992.32	1927	6	5.97 %
30b Mechanical calculators and typewriters	2572.21	1971	2605.83	1972	2647.87	1973	2689.90	1974	2723.53	1976	5	5.88 %
30c Mechanical calculators and typewriters	2674.32	1980	2907.04	1985	3197.95	1988	3488.85	1991	3721.58	1996	16	39.16 %
33 Telecommunications	5157.97	1919	6151.88	1926	7394.25	1930	8636.63	1935	9630.53	1941	22	86.71 %
34 Other electrical communication systems	2093.30	1928	2276.23	1934	2504.90	1937	2733.56	1941	2916.50	1946	18	39.33 %
35a Special radio systems	463.28	1928	1883.58	1947	3658.95	1958	5434.33	1969	6854.63	1988	60	1379.59 %
35b Special radio systems	5719.06	1986	5780.75	1987	5857.87	1988	5934.98	1990	5996.67	1991	5	4.85 %
36a Image and sound equipment	5469.15	1924	5841.90	1927	6307.83	1929	6773.77	1931	7146.51	1933	9	30.67 %
36b Image and sound equipment	6302.64	1961	10555.30	1977	15871.20	1986	21187.00	1995	25439.70	2010	49	303.64 %
37a Illumination devices	7119.99	1918	8261.84	1926	9689.15	1930	11116.50	1935	12258.30	1942	24	72.17 %

Table 4.3. (cont.)

37b	Illumination devices	10708.80	1956	10977.00	1957	11312.20	1958	11647.40	1959	11915.50	1960	4	11.27 %
38a	Electrical devices and systems	8140.86	1910	13366.20	1922	19897.80	1929	26429.50	1936	31654.80	1948	38	288.84 %
38b	Electrical devices and systems	28029.60	1948	28472.60	1949	29026.30	1950	29580.10	1951	30023.10	1952	4	7.11 %
38c	Electrical devices and systems	30966.10	1956	36214.70	1963	42775.40	1968	49336.10	1972	54584.70	1980	24	76.27 %
38d	Electrical devices and systems	50262.80	1985	51151.70	1987	52262.70	1988	53373.80	1989	54262.70	1991	6	7.96 %
39a	Other general electrical equipment	12890.20	1919	14893.90	1926	17398.40	1930	19902.90	1934	21906.50	1940	21	69.95 %
39b	Other general electrical equipment	20612.10	1948	20900.40	1949	21260.80	1949	21621.10	1950	21909.40	1950	2	6.29 %
39c	Other general electrical equipment	22776.30	1957	27013.30	1966	32309.60	1972	37605.90	1978	41842.90	1987	30	83.71 %
40	Semiconductors	510.50	1955	2379.97	1963	4716.81	1968	7053.65	1973	8923.12	1981	26	1647.92 %
41	Office equipment and data processing systems	5197.61	1957	15191.90	1974	27684.80	1985	40177.70	1995	50172.00	2013	56	865.29 %
52a	Photographic equipment	2022.60	1919	2044.61	1921	2072.13	1923	2099.64	1925	2121.65	1928	9	4.90 %
52b	Photographic equipment	2546.62	1948	2592.96	1949	2650.88	1950	2708.80	1950	2755.14	1951	3	8.19 %
52c	Photographic equipment	2436.16	1955	4760.59	1968	7666.14	1976	10571.70	1984	12896.10	1997	42	429.36 %

*, **, ***: *See footnote for Table 4.2*

Table 4.4. Cycle properties: Mechanical

Code**	Technological group	Socio-economic capability at: P(t5%)*	Year	Socio-economic capability at: P(t25%)	Takeoff year	Socio-economic capability at: P(t50%)	Year	Socio-economic capability at: P(t75%)	Year	Socio-economic capability at: P(t95%)	Year	Cycle duration in years	Cycle opportunity***: (Growth rate)
		Depression *				Prosperity							
				Revival				Recession					
1a	Food and tobacco products	3242.87	1922	3769.34	1929	4427.42	1934	5085.50	1938	5611.97	1945	23	73.06 %
1b	Food and tobacco products	5339.40	1959	6194.82	1966	7264.09	1971	8333.36	1975	9188.78	1983	24	72.09 %
1c	Food and tobacco products	9123.89	1988	9271.75	1988	9456.57	1989	9641.38	1989	9789.24	1990	2	7.29 %
13a	Metallurgical processes	5613.79	1916	7197.26	1925	9176.61	1930	11156.00	1935	12739.42	1944	28	126.93 %
13b	Metallurgical processes	10360.20	1958	13291.30	1966	16955.13	1970	20619.00	1975	23550.02	1982	24	127.31 %
14a	Miscellaneous metal products	76082.20	1915	79419.10	1922	83590.27	1926	87761.40	1930	91098.36	1937	22	19.74 %
14b	Miscellaneous metal products	64675.30	1986	66335.70	1988	68411.22	1989	70486.70	1990	72147.15	1992	6	11.55 %
15a	Food. Drink and tobacco equipment	2905.69	1950	2943.09	1950	2989.84	1951	3036.59	1951	3073.99	1952	2	5.79 %
15b	Food. Drink and tobacco equipment	1955.10	1955	1956.07	1956	1957.27	1957	1958.48	1958	1959.45	1960	5	0.22 %

Table 4.4. (cont.)

15c	Food. Drink and tobacco equipment	2901.69	1964	3042.89	1967	3219.39	1969	3395.89	1972	3537.09	1975	11	21.90 %
16a	Chemical and allied equipment	25115.90	1921	26720.80	1926	28726.96	1929	30733.10	1933	32338.06	1938	17	28.76 %
16b	Chemical and allied equipment	26819.70	1950	26944.70	1951	27100.91	1951	27257.20	1952	27382.16	1952	2	2.10 %
16c	Chemical and allied equipment	27799.60	1961	30815.50	1967	34585.36	1971	38355.20	1974	41371.14	1981	20	48.82 %
17a	Metal working equipment	25555.10	1963	27595.10	1968	30145.09	1970	32695.10	1973	34735.09	1978	15	35.92 %
17b	Metal working equipment	32005.10	1984	32173.20	1985	32383.45	1986	32593.70	1987	32761.86	1989	5	2.36 %
18a	Paper making apparatus	6257.36	1923	6961.61	1928	7841.91	1931	8722.21	1934	9426.45	1938	15	50.65 %
18b	Paper making apparatus	7738.91	1957	8823.71	1964	10179.72	1968	11535.70	1972	12620.53	1979	22	63.08 %
19a	Building material processing equipment	4203.81	1920	4375.62	1923	4590.38	1925	4805.13	1927	4976.94	1931	11	18.39 %
19b	Building material processing equipment	2882.51	1962	3054.75	1966	3270.05	1969	3485.35	1971	3657.59	1976	14	26.89 %
20	Assembly and material handling equipment	19374.30	1950	23515.10	1961	28690.95	1967	33866.90	1974	38007.58	1984	34	96.18 %
21a	Agricultural equipment	5705.55	1947	5899.76	1952	6142.52	1956	6385.28	1959	6579.48	1964	17	15.32 %
21b	Agricultural equipment	6603.58	1968	6733.28	1969	6895.40	1970	7057.52	1971	7187.21	1972	4	8.84 %
22a	Other construction and excavating equipment	2520.50	1922	2574.21	1923	2641.35	1924	2708.48	1925	2762.19	1926	4	9.59 %
22b	Other construction and excavating equipment	2722.02	1928	2766.11	1930	2821.21	1931	2876.32	1932	2920.40	1934	6	7.29 %

Table 4.4. (cont.)

22c	Other construction and excavating equipment	1804.40	1958	1934.24	1964	2096.53	1968	2258.83	1971	2388.66	1977	19	32.38 %
23a	Mining equipment	1952.67	1919	2603.99	1926	3418.14	1931	4232.28	1935	4883.60	1942	23	150.10 %
23b	Mining equipment	4186.17	1954	5288.75	1962	6666.98	1966	8045.21	1971	9147.79	1979	25	118.52 %
24a	Electrical lamp manufacturing	83.61	1924	126.07	1929	179.15	1932	232.22	1935	274.68	1940	16	228.53 %
24b	Electrical lamp manufacturing	358.53	1959	435.12	1966	530.86	1969	626.60	1973	703.20	1979	20	96.13 %
25	Textile and clothing machinery	17997.40	1921	18788.80	1926	19778.17	1930	20767.50	1933	21558.99	1938	17	19.79 %
26a	Printing and publishing machinery	4957.13	1964	5015.73	1965	5088.98	1965	5162.23	1966	5220.83	1966	2	5.32 %
26b	Printing and publishing machinery	5141.90	1969	5245.50	1972	5375.00	1973	5504.50	1975	5608.10	1978	9	9.07 %
27a	Woodworking tools and machinery	943.54	1949	1013.69	1953	1101.39	1955	1189.08	1958	1259.24	1962	13	33.46 %
27b	Woodworking tools and machinery	1261.22	1963	1325.75	1968	1406.43	1971	1487.10	1974	1551.64	1979	16	23.03 %
28a	Other specialised machinery	32208.40	1919	33904.60	1924	36024.82	1927	38145.10	1930	39841.29	1935	16	23.70 %
28b	Other specialised machinery	28911.60	1950	29046.00	1951	29214.04	1951	29382.10	1952	29516.48	1953	3	2.09 %
28c	Other specialised machinery	29220.80	1956	31832.20	1963	35096.42	1967	38360.70	1971	40972.09	1979	23	40.22 %

Table 4.4. (cont.)

29a	Other general industrial equipment	57679.00	1921	60275.90	1926	63522.00	1929	66768.10	1931	69364.99	1936	15	20.26 %
29b	Other general industrial equipment	48381.90	1961	51296.90	1966	54940.65	1969	58584.40	1972	61499.42	1976	15	27.11 %
29c	Other general industrial equipment	60835.70	1984	61295.50	1986	61870.31	1988	62445.10	1989	62904.90	1992	8	3.40 %
31	Power plants	2377.85	1944	3632.17	1955	5200.06	1962	6767.96	1968	8022.27	1979	35	237.37 %
50a	Non-metallic mineral products	11841.90	1920	13362.80	1927	15263.85	1931	17165.00	1935	18685.84	1941	21	57.79 %
50b	Non-metallic mineral products	13654.10	1958	18231.40	1969	23953.01	1975	29674.60	1981	34251.89	1992	34	150.85 %
53a	Other instruments and controls	35016.30	1918	37063.20	1924	39621.74	1927	42180.30	1931	44227.19	1936	18	26.30 %
53b	Other instruments and controls	37880.90	1946	56444.60	1966	79649.09	1978	102853.62	1990	121417.30	2009	63	220.52 %

*. **. ***: *See footnote for Table 4.2*

Table 4.5. Cycle properties: Transport

Code**	Technological group	Depression *				Prosperity							
				Revival				Recession					
		Socio-economic capability at: P(t5%)*	Year	Socio-economic capability at: P(t25%)	Takeoff year	Socio-economic capability at: P(t50%)	Year	Socio-economic Capability at: P(t75%)	Year	Socio-economic capability at: P(t95%)	Year	Cycle duration in years	Cycle opportunity***: (Growth rate)
42	Internal combustion engines	4914.41	1970	7012.02	1978	9634.02	1983	12256.00	1988	14353.60	1997	27	192.07 %
43	Motor vehicles	3107.28	1954	4056.74	1966	5243.58	1972	6430.40	1979	7379.88	1991	37	137.50 %
44	Aircraft	2690.01	1923	3240.01	1927	3927.51	1930	4615.00	1933	5165.01	1938	15	92.01 %
45a	Ships and marine propulsion	3726.74	1929	3777.05	1930	3839.94	1930	3902.80	1931	3953.14	1932	3	6.08 %
45b	Ships and marine propulsion	2219.79	1956	2936.47	1963	3832.32	1968	4728.20	1972	5444.85	1979	23	145.29 %
46	Railways and railway equipment	4506.33	1969	4669.33	1971	4873.08	1973	5076.80	1974	5239.83	1976	7	16.28 %
49a	Rubber and plastic products	7115.35	1919	7376.53	1920	7703.02	1921	8029.50	1922	8290.69	1924	5	16.52 %
49b	Rubber and plastic products	4173.85	1952	5616.16	1963	7419.06	1969	9222.00	1976	10664.30	1987	35	155.50 %

*, **, ***: *See footnote for Table 4.2*

Table 4.6. Cycle properties: Non-industrial

Code**	Technological group	Depression *				Prosperity						Cycle duration in years	Cycle opportunity***: (Growth rate)
				Revival				Recession					
		Socio-economic capability at: P(t5%)*	Year	Socio-economic Capability at: P(t25%)	Takeoff year	Socio-economic Capability at: P(t50%)	Year	Socio-economic Capability at: P(t75%)	Year	Socio-economic capability at: P(t95%)	Year		
32a	Nuclear reactors	127.31	1956	636.54	1961	1273.08	1965	1909.61	1968	2418.84	1974	18	1799.96 %
32b	Nuclear reactors	2773.45	1985	2867.88	1986	2985.93	1987	3103.97	1987	3198.40	1988	3	15.32 %
48	Textiles, clothing and leather	7767.33	1917	8118.79	1922	8558.12	1925	8997.44	1928	9348.90	1934	17	20.36 %
54a	Wood products	4636.08	1949	4768.68	1950	4934.43	1951	5100.18	1952	5232.78	1953	4	12.87 %
54b	Wood products	5298.74	1957	5477.03	1961	5699.89	1964	5922.76	1966	6101.05	1971	14	15.14 %
54c	Wood products	5924.46	1970	5986.46	1972	6063.96	1972	6141.46	1973	6203.46	1975	5	4.71 %
54d	Wood products	5419.52	1984	5615.63	1987	5860.78	1988	6105.92	1989	6302.04	1991	7	16.28 %
56a	Other manufacturing and non-industrial	27945.10	1920	28791.60	1923	29849.80	1924	30908.10	1926	31754.60	1929	9	13.63 %
56b	Other manufacturing and non-industrial	23358.40	1961	25789.00	1967	28827.30	1970	31865.70	1974	34296.30	1980	19	46.83 %

*. **. ***: *See footnote for Table 4.2*

However, there is no evidence for innovation cycles to be longer or shorter depending on the historical period. Given no fixed periodicity of the time-span of each cycle, as well as no evidence for innovation cycles to be longer or shorter depending on the historical period, these results are in accordance with the findings of Schumpeter and Kuznets, who saw no need for such cycles to have a fixed length as the nature and impact of each new innovation is different.[11] That is, the length of the cycle depends on the nature and environment of the particular innovation (Schumpeter 1939, p. 119).

The results also indicate cross-technology differences in terms of the size of accumulated patent stocks at different stages of development, reflecting different socio-economic capability or historical impact, and in terms of change or growth in accumulated patent stock across the full cycle duration, reflecting different opportunities across technologies. These results certainly confirm the diversity of technology dynamics in terms of cyclical phases, cycle duration, technological opportunity and level of socio-economic capability or impact.

4.4.3. Clustering of Innovations: A Brief Exploration

As each cycle does not have the same time span, the cyclical phases are not appropriate to focus on when comparing the timing of innovation cycles. What is important is the point of technological takeoff. Synchronisation of such has often been used in tracing possible clusters of innovation. The year of technological takeoff for each growth cycle presented in Tables 4.2-4.6 has been plotted in Graph 4.45.

It is important to note that periods with no takeoff in technological patent groups or trajectories are not indicative of a lack of inventive or innovative activity. The cycle might be on the inflection point with the most rapid rate of development, or may be well into the innovative prosperity phase. It may also be a period when radical (or basic) inventions are being patented, and individual patents are very significant for future exploitation, but are not important in numbers (i.e. the field as a whole has not reached the collective competence necessary for takeoff or revival).

From Graph 4.45 it can be observed that although takeoffs do not move in any synchronised fashion, they do seem to follow a certain band-like (or epoch-like) character, with a clustering of takeoffs from 1920 (the beginning of the data set eligible for analysis) to the early 1930s. Then there is a gap with no single takeoff up until the late 1940s, followed by numerous takeoffs (especially in the upswing of the 1960s) up to about the early 1970s. Hence, Graph 4.45

---

[11] The 'cycle duration' issue (i.e. time span of each S-curve) discussed here ought not to be confused with the long 'wave' debate with is about 'clustering of innovations' (or synchronisation of cycle takeoffs).

does suggest that something puzzling is happening with the takeoff bands. (However, the fact that takeoffs are not close to 1990, is presumably a reflection on the constraints of the data; i.e. that the technological growth curve has to be at least 60% in 1990 to make sure that we are dealing with an S-curve). There could be various (theoretical competing) explanations for this, and they are set out and discussed in Chapter 5.

Graph 4.45. Timing of technological cycle takeoff [technological group; year]*

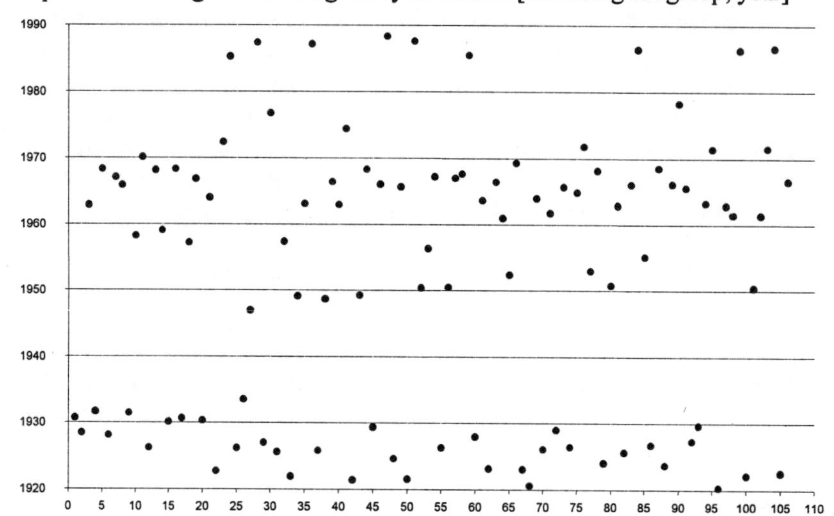

* *Technological trajectories are randomly numbered on the x-axis (i.e. they do not reflect the technological codes in Tables 4.2-4.6): Chemicals (1-21), Electrical/electronics (22-44), Mechanical (56-89), Transport (90-97) and Non-industrial (98-106).*

Finally, grouping the technological growth cycles with respect to their bands of takeoffs within this century (as identified in the text above, and in Graph 4.45), and sorting those technologies within each band by cycle opportunity (cycle growth rates) and historical impact or socio-economic capability (accumulated patent stocks), (see Tables 4.2-4.6), it can be observed that epochs differ radically with respect to (i) the technologies involved, including (ii) their scope (i.e. socie-economic capability (or accumulated patent stock) and opportunity (or cycle growth rate)) and (iii) overall 'technological configuration' (i.e. the combined structure of (i)and (ii)).

Assuming that the overall configuration of technologies within a certain band of takeoff can be associated with an overall system takeoff, this suggests how we might start to measure and understand evolving technological systems of innovation clusters. Such viewpoints are carried further in Chapter 5, which analyses clusters of innovative activity and the cyclical theories of waves in a system perspective.

4.5. CONCLUSION

Using Kuznets's cyclical units of troughs and peaks as representing the growth cycle of a technology, and applying them to Schumpeter's four phases of development (depression, revival, prosperity and recession), it can be concluded that technological growth tends to follow an *S*-shaped growth curve, and that the biological growth model in logistic form is an appropriate approximation to describe the paths of evolution.

When identifying the technology cycles of sustained growth (as opposed to periods of random fluctuations), we saw how we often have periods of punctuated equilibrium in which one growth curve follows another with no period of absolute decline. However, periods of sustained decline (i.e. crisis), when they occur, tend to be of very short duration in comparison to the growth cycle. This suggests that the *S*-shaped growth curve does not reflect only half of a technology's 'life', as has sometimes been argued.

The results also indicate that, although the timing of takeoffs tends to be clustered within certain time bands (normally associated with upswing periods), the technologies vary in terms of the timing of operation (i.e. no synchronisation in the timing of phases of depression, revival, prosperity and recession) due to the great diversity in cycle duration. The results also indicate cross-technology differences in terms of collective experience and opportunity (identified from the accumulated patent stock and change or growth in stock over the life cycle, respectively). This confirms the variety of technology dynamics, as well as a scope for how we might start to understand evolving waves of innovations within clusters or technology systems (an issue which is addressed further in Chapter 5).

The results also show that most of the *S*-shaped growth cycles are not subject to significant external shocks, but are carried through or survive almost all the way to their potential long-run ceiling level. This reinforces the relevance of Schumpeter's point that development proceeds not in a smooth line but in cycles, and that although external factors have a large influence on change and growth, it is still very important to understand the mechanics and dynamics internal to the system.

4.6. APPENDIX

Table 4.7. Chemical technological growth-periods between turning points

Code	Groups of 56 related technological fields	Sustained growth periods from trough to peak			
2	Distillation processes	1920-1983			
3	Inorganic chemicals	1920-1940	1950-1952	1955-1985	
4	Agricultural chemicals	1920-1927	1930-1942	1950-1990	
5	Chemical processes	1920-1944	1949-1990		
6	Photographic chemistry	1920-1990			
7	Cleaning agents and other compositions	1920-1946	1949-1952	1955-1962	1964-1990
8	Disinfecting and preserving	1920-1940	1942-1945	1948-1953	1955-1990
9	Synthetic resins and fibres	1920-1990			
10	Bleaching and dyeing	1920-1942	1958-1983		
11	Other organic compounds	1920-1978			
12	Pharmaceuticals and biotechnology	1920-1990			
51	Coal and petroleum products	1920-1949	1956-1960	1964-1977	1982-1985
55	Explosive compositions and charges	1920-1923	1928-1934	1937-1940	1959-1978

Periods starting in 1920 or ending in 1990 may reflect the time-limit of the database used for the analysis rather than indicating troughs and peaks. Periods with less than five observations are shown in grey, and they are be disregarded in the statistical analysis. Growth periods are underlined if they are linked by punctuated equilibrium (assuming they can be fitted to S-shaped curves).

Table 4.8. Electrical/electronics technological growth periods between turning points

Code/Groups of 56 related technological fields	Sustained growth periods from trough to peak					
30 Mechanical calculators and typewriters	1921–1926	1949–1951	1957–1959	1965–1967	1971–1976	1979–1990
33 Telecom-munications	1920–1942	1949–1952	1957–1962	1964–1990		
34 Other electrical communication systems	1923–1944	1947–1962				
35 Special radio systems	1920–1978	1986–1990				
36 Image and sound equipment	1924–1932	1938–1941	1949–1951	1955–1990		
37 Illumination devices	1920–1942	1948–1952	1955–1962	1964–1975	1986–1990	
38 Electrical devices and systems	1920–1942	1948–1953	1955–1977	1983–1990		
39 Other general electrical equipment	1920–1942	1948–1952	1955–1990			
40 Semiconductors	1920–1981	1984–1990				
41 Office equipment and data processing systems	1920–1942	1948–1990				
52 Photographic equipment	1920–1926	1930–1944	1948–1952	1955–1990		

Note: See footnote to Table 4. 7.

Table 4.9. Mechanical technological growth periods between turning points

Code / Groups of 56 related technological fields	Sustained growth periods from trough to peak				
1 Food and tobacco products	1920–1943	1949–1952	1955–1981	1985–1990	
13 Metallurgical processes	1920–1942	1955–1990			
14 Miscellaneous metal products	1920–1933	1950–1952	1957–1959	1963–1976	1985–1990
15 Food, drink and tobacco equipment	1920–1941	1949–1953	1955–1959	1964–1976	1986–1990
16 Chemical and allied equipment	1920–1935	1949–1953	1955–1990		
17 Metal working equipment	1920–1934	1955–1959	1961–1977	1984–1990	
18 Paper making apparatus	1920–1940	1957–1976	1988–1990		
19 Building material processing equipment	1920–1933	1957–1959	1963–1977		
20 Assembly and material handling equipment	1920–1933	1949–1977	1987–1990		
21 Agricultural equipment	1949–1965	1968–1973			
22 Other construction and excavating equipment	1920–1926	1928–1932	1957–1977		
23 Mining equipment	1920–1944	1951–1990			
24 Electrical lamp manufacturing	1920–1943	1950–1952	1955–1978	1988–1990	
25 Textile and clothing machinery	1920–1941	1965–1975			
26 Printing and publishing machinery	1923–1934	1963–1967	1970–1977	1982–1990	

27	Woodworking tools and machinery	1926–1931	1950–1962	1964–1978	1980–1990
28	Other specialised machinery	1920–1934	1949–1953	1955–1976	
29	Other general industrial equipment	1920–1934	1959–1981	1984–1990	
31	Power plants	1930–1980	1986–1989		
50	Non-metallic mineral products	1920–1942	1957–1990		
53	Other instruments and controls	1920–1936	1938–1941	1948–1990	

Note: See footnote to Table 4.7.

Table 4.10. Transport technological growth-periods between turning points

Code / Groups of 56 related technological fields		Sustained growth periods from trough to peak				
42	Internal combustion engines	1920-1934	1957-1959	1964-1990		
43	Motor vehicles	1920-1932	1955-1990			
44	Aircraft	1920-1933	1944-1946	1948-1951	1959-1973	1986-1988
45	Ships and marine propulsion	1928-1932	1949-1951	1954-1978	1984-1990	
46	Railways and railway equipment	1965-1975				
47	Other transport equipment	1920-1926	1957-1962	1965-1978	1985-1990	
49	Rubber and plastic products	1920-1926	1950-1952	1955-1985	1987-1990	

Note: See footnote to Table 4.7.

Table 4.11. Non-industrial technological growth-periods between turning points

Code / Groups of 56 related technological fields		Sustained growth periods from trough to peak				
32	Nuclear reactors	1938-1978	1980-1982	1985-1990		
48	Textiles, clothing and leather	1920-1932	1984-1990			
54	Wood products	1920-1922	1949-1953	1955-1967	1970-1975	1984-1990
56	Other manufacturing and non-industrial	1920-1932	1949-1952	1955-1981	1983-1990	

Note: See footnote to Table 4.7.

Table 4.12. Parameter results: Chemicals

Code*	Technological group	Period used for model fitting: t_0=first obs.	P(t initial) K	Ceiling $\bar{p}$	P(t ceiling) $\bar{p}$+K	$\hat{a}$	$\hat{b}$	R^{2**}
2	Distillation processes	1920–83	3.80	1808.84	1812.64	–1.74	0.06	0.9928
3a	Inorganic chemicals	1920–40	2661.59	1982.33	4643.93	–4.41	0.39	0.9819 **
3b	Inorganic chemicals	1955–85	3609.53	5811.04	9420.58	–2.76	0.21	0.9948
4a	Agricultural chemicals	1930–42	310.13	163.30	473.43	–2.56	0.89	0.9825
4b	Agricultural chemicals	1950–90	396.37	4459.79	4856.16	–3.43	0.13	0.9947 **
5a	Chemical processes	1920–44	4120.37	7124.13	11244.50	–3.15	0.25	0.9972 *
5b	Chemical processes	1949–90	10361.33	29360.8	39722.11	–4.21	0.17	0.9954 **
6	Photographic chemistry	1920–90	385.17	14614.7	14999.83	–5.51	0.10	0.9816 **
7a	Cleaning agents and other compositions	1920–46	3154.13	6875.61	10029.74	–3.24	0.19	0.9978 *
7b	Cleaning agents and other compositions	1955–62	8900.00	1604.89	10504.89	–19.91	0.49	0.9907
7c	Cleaning agents and other compositions	1964–90	9904.47	7118.70	17023.17	–2.63	0.25	0.9847
8a	Disinfecting and preserving	1920–40	36.63	72.36	109.00	–2.17	0.17	0.9656
8b	Disinfecting and preserving	1955–90	99.60	452.99	552.59	–2.91	0.14	0.9856
9	Synthetic resins and fibres	1920–90	537.47	43804.80	44342.27	–4.72	0.09	0.9950 **
10a	Bleaching and dyeing	1920–42	632.77	1849.94	2482.70	–4.00	0.29	0.9974 **
10b	Bleaching and dyeing	1958–83	1664.13	1885.04	3549.18	–4.58	0.34	0.9972 **
51a	Coal and petroleum products	1920–49	815.77	7412.66	8228.43	–3.56	0.23	0.9972 *
51b	Coal and petroleum products	1956–60	7637.95	973.09	8611.04	–3.53	2.02	0.9993 *
51c	Coal and petroleum products	1964–77	8466.17	1883.87	10350.04	–2.51	0.50	0.9722 *
55a	Explosive compositions and charges	1928–34	408.58	73.28	481.86	–3.48	1.02	0.9946 *
55b	Explosive compositions land charges	1959–78	359.97	1101.06	1461.03	–1.11	0.26	0.9756

** a, b, and c, after the technological group code number, denote different cycles of same technological group across different periods in time. (See Tables 4.7. to 4. 11)*

*** The first empirical observation is below 2% (**) or 5% (*) of the growth cycle's ceiling level. Otherwise it is between 5% and 10% (or perhaps between 5% and 25% if t_0 is 1920).*

Table 4.13. Parameter results: Electrical / electronics

Code*	Technological group	Period used for model fitting: t_0=first obs.	P(t initial) K	Ceiling $\bar{p}$	P(t ceiling) $\bar{p}$+K	$\hat{a}$	$\hat{b}$	R^{2**}
30a	Mechanical calculators and typewriters	1921–26	4695.47	312.48	5007.95	–2.97	1.06	0.9999 *
30b	Mechanical calculators and typewriters	1971–77	2563.80	168.14	2731.94	–2.93	1.29	0.9923
30c	Mechanical calculators and typewriters	1979–90	2616.14	1163.62	3779.76	–3.37	0.36	0.9981 *
33	Telecommunications	1920–42	4909.50	4969.50	9879.00	–2.74	0.27	0.9797
34	Other electrical communication systems	1923–44	2047.57	914.66	2962.23	–4.41	0.31	0.9933 **
35a	Special radio systems	1920–78	108.20	7101.51	7209.71	–3.75	0.10	0.9952 *
35b	Special radio systems	1986–90	5703.64	308.46	6012.10	–2.51	1 02	0.9587
36a	Image and sound equipment	1924–32	5375.97	1863.73	7239.70	–2.99	0.63	0.9944 *
36b	Image and sound equipment	1955–90	5239.47	21263.4	26502.84	–3.73	0.12	09858 *
37a	Illumination devices	1920–42	6834.53	5709.23	12543.76	–2.47	0.24	0.9855
37b	Illumination devices	1955–62	10641.78	1340.80	11982.58	–4.76	1.54	0.9931 **
38a	Electrical devices and systems	1920–42	6834.53	26126.6	32961.12	–1.40	0.15	0.9917
38b	Electrical devices and systems	1948–53	27918.80	2215.08	30133.88	–2.80	1.51	0.9990
38c	Electrical devices and systems	1955–77	29653.98	26242.80	55896.79	–3.09	0.24	0.9920 *
38d	Electrical devices and systems	1983–90	50040.57	4444.32	54484.89	–5.19	0.98	0.9912 **
39a	Other general electrical equipment	1920–42	12389.33	10018.10	22407.43	–2.70	0.28	0.9923
39b	Other general electrical equipment	1948–52	20540.03	1441.48	21981.51	–2.90	2.54	0.9932
39c	Other general electrical equipment	1955–90	21717.03	21185.1	42902.12	–3.30	0.19	0.9818 *
40	Semiconductors	1920–81	43.13	9347.36	9390.49	–10.92	0.23	0.9988 **
41	Office equipment and data processing systems	1948–90	2699.03	49971.50	52670.53	–3.86	0.10	0.9784 *
52a	Photographic equipment	1920–26	2017.10	110.06	2127.16	–2.02	0.66	0.9491
52b	Photographic equipment	1948–52	2535.03	231.69	2766.72	–3.74	2.11	0.9992 *
52c	Photographic equipment	1955–90	1855.05	11622.20	13477.25	–2.96	0.14	0.9958 *

* ** *See footnote to Table 4.12.*

Table 4.14. Parameter results: Mechanical

Code*	Technological group	Period used for model fitting: t_0=first obs.	P(t initial) K	Ceiling $\bar{p}$	P(t ceiling) $\bar{p}$+K	$\hat{a}$	$\hat{b}$	R^{2**}
la	Food and tobacco products	1920–43	3111.25	2632.33	5743.59	–3.47	0 25	0.9900 *
lb	Food and tobacco products	1955–81	5125.55	4277.08	9402.63	–3.81	0.24	0.9906 *
lc	Food and tobacco products	1985–90	9086.93	739.27	9826.20	–8.60	2.25	0.9992 **
13a	Metallurgical processes	1920–42	5217.92	7917.37	13135.29	–2.07	0.21	0.9972
13b	Metallurgical processes	1955–90	9627.47	14655.30	24282.78	–12.36	0.25	0.9982 **
14a	Miscellaneous metal products	1920–33	75247.94	16684.70	91932.60	–1.52	0.27	0.9919
14b	Miscellaneous metal products	1985–90	64260.19	8302.06	72562.25	–3.51	0.93	0.9925 *
15a	Food. Drink and tobacco equipment	1949–53	2896.34	187.00	3083.34	–4.07	2.10	0.9998 **
15b	Food. Drink and tobacco equipment	1955–59	1954.86	4.83	1959.69	–2.56	1.10	0.9865
15c	Food. Drink and tobacco equipment	1964–76	2866.39	706.00	3572.39	–2.74	0.51	0.9866
16a	Chemical and allied equipment	1920–35	24714.62	8024.67	32739.29	–3.26	0.34	0.9906 *
16b	Chemical and allied equipment	1949–53	26788.41	625.00	27413.41	–4.26	2.07	0.9993 **
16c	Chemical and allied equipment	1955–90	27045.61	15079.50	42125.12	–4.65	0.30	0.9967 **
17a	Metal working equipment	1961–77	25045.09	10200.00	35245.09	–3.80	0.41	0.9851 *
17b	Metal working equipment	1984–90	31963.00	840.90	32803.90	–2.93	1.23	0.9450
18a	Paper making apparatus	1920–40	6081.30	3521.21	9602.52	–4.24	0.39	0.9956 **
18b	Paper making apparatus	1957–76	7467.71	5424.02	12891.73	–2.83	0.26	0.9919
19a	Building material processing equipment	1920–33	4160.86	859.03	5019.89	–2.71	0.51	0.9703
19b	Building material processing equipment	1963–77	2839.45	861.20	3700.65	–2.60	0.44	0.9686
20	Assembly and material handling equipment	1949–77	18339.14	20703.60	39042.77	–3.19	0.17	0.9969 *
21a	Agricultural equipment	1949–65	5657.00	971.04	6628.04	–2.29	0.35	0.9812
21b	Agricultural equipment	1968–73	6571.15	648.48	7219.64	–3.14	1.55	0.9928 *
22a	Other construction and excavating equipment	1920–26	2507.08	268.54	2775.62	–5.10	1.30	0.9844 **

22b Other construction and excavating equipment	1928–32	2711.00	220.42	2931.42	–2.75	1.03	0.9786	
22c Other construction and excavating equipment	1957–77	1771.94	649.18	2421.12	–3.22	0.30	0.9868	*
23a Mining equipment	1920–44	1789.84	3256.59	5046.43	–2.60	0.25	0.9919	
23b Mining equipment	1951–90	3910.53	5512.91	9423.44	–3.62	0.23	0.9936	*
24a Electrical lamp manufacturing	1920–43	73.00	212.29	285.29	–4.39	0.37	0.9935	*
24b Electrical lamp manufacturing	1955–78	339.38	382.97	722.35	–4.24	0.30	0.9755	*
25 Textile and clothing machinery	1920–41	17799.48	3957.37	21756.86	–3.32	0.35	0.9955	*
26a Printing and publishing machinery	1963–67	4942.48	293.00	5235.48	–5.90	2.57	0.9997	**
26b Printing and publishing machinery	1970–77	5116.00	518.00	5634.00	–2.37	0.71	0.9768	
27a Woodworking tools and machinery	1950–62	926.00	350.78	1276.78	–2.48	0.46	0.9519	
27b Woodworking tools and machinery	1964–78	1245.08	322.70	1567.78	–2.60	0.36	0.9747	
28a Other specialised machinery	1920–34	31784.30	8481.04	40265.34	–2.57	0.37	0.9876	
28b Other specialised machinery	1949–53	28878.00	672.08	29550.08	–4.62	1.95	0.9947	**
28c Other specialised machinery	1955–76	28567.90	13057.00	41624.94	–3.14	0.26	0.9902	*
29a Other general industrial equipment	1920–34	57029.79	12984.40	70014.21	–3.24	0.38	0.9860	*
29b Other general industrial equipment	1959–81	47653.12	14575.10	62228.18	–3.81	0.39	0.9897	*
29c Other general industrial equipment	1984–90	60720.76	2299.09	63019.86	–2.85	0.73	0.9617	
31 Power plants	1930–80	2064.27	6271.58	8335.85	–5.35	0.17	0.9965	**
50a Non-metallic mineral products	1920–42	11461.65	7604.41	19066.06	–2.98	0.28	0.9990	*
50b Non-metallic mineral products	1957–90	12509.80	22886.40	35396.22	–3.11	0.17	0.9807	*
53a Other instruments and controls	1920–36	34504.57	10234.30	44738.91	–2.26	0.32	0.9860	
53b Other instruments and controls	1948–90	33240.02	92818.10	126058.15	–2.79	0.09	0.9840	

* ** *See footnote to Table 4.12.*

Table 4.15. Parameter results: Transport

Code*	Technological group	Period used for model fitting: t_0=first obs.	P(t initial) K	Ceiling $\bar{p}$	P(t ceiling) $\bar{p}$+K	$\hat{a}$	$\hat{b}$	R^{2**}
42	Internal combustion engines	1964–90	4390.01	10488.00	14878.03	–4.23	0.22	0.9976 **
43	Motor vehicles	1955–90	2869.91	4747.34	7617.25	–2.79	0.16	0.9777
44	Aircraft	1920–33	2552.51	2750.00	5302.51	–3.94	0:39	0.9693 **
45a	Ships and marine propulsion	1928–32	3714.16	251.56	3965.72	–3.98	1.75	0.9881 **
45b	Ships and marine propulsion	1954–78	2040.62	3583.40	5624.01	–3.51	0.26	0.9887 *
46	Railways and railway equipment	1965–75	4465.58	815.00	5280.58	–6.79	0.89	0.9831 **
49a	Rubber and plastic products	1920–26	7050.05	1305.94	8355.98	–1.46	1.12	0.9968
49b	Rubber and plastic products	1955–85	3813.27	7211.58	11024.85	–2.43	0.17	0.9982

* ** *See footnote to Table 4.12.*

Table 4.16. Parameter results: Non-industrial

Code*	Technological group	Period used for model fitting: t_0=first obs.	P(t initial) K	Ceiling $\bar{p}$	P(t ceiling) $\bar{p}$+K	$\hat{a}$	$\hat{b}$	R^{2**}
32a	Nuclear reactors	1938–78	0.00	2546.15	2546.15	–8.99	0.34	0.9955 **
32b	Nuclear reactors	1985–90	2749.84	472.16	3222.01	–3.86	2.17	0.9928 *
48	Textiles, clothing and leather	1920–32	7679.46	1757.30	9436.77	–1.86	0.35	0.9921
54a	Wood products	1949–53	4602.93	663.00	5265.93	–3.02	1.33	0.9909 *
54b	Wood products	1955–67	5254.16	891.46	6145.63	–3.89	0.44	0.9593 *
54c	Wood products	1970–75	5908.96	310.00	6218.96	–2.81	1.13	0.9631
54d	Wood products	1984–90	5370.49	980.58	6351.07	–3.33	0.86	0.9933 *
56a	Other manufacturing and non-industrial	1920–32	27733.43	4232.82	31966.25	–2.82	0.67	0.9805
56b	Other manufacturing and non-industrial	1955–81	22750.70	12153.3	34903.98	-4.68	0.31	0.9906 **

* ** *See footnote to Table 4 12.*

5. Clusters of Takeoffs in Innovation Trajectories: An Exploration of Wave-like Patterns

Abstract

This chapter explores wave-like patterns as a phenomenon in real quantities. Special attention is given to analysing periodic clustering of established innovation trajectories (rather than of 'basic' or radical innovations). Due to a demonstrated great diversity in the time span of the cyclical phases of each of the established innovation trajectories, only the clustering of trajectory-takeoffs at the technological frontier is the focus of attention.

Cyclical theories of innovative activity are linked with a system perspective in order to derive the changing population of innovation trajectories participating in different innovation waves. By use of panel data analysis, the internal composition and changing technological scope of such changing populations are also examined.

The results confirm wave-like patterns in technological activity at the technological frontier, and the internal composition and technological scope of such waves is comparable with the third, fourth and an emerging fifth Kondratiev (1925).

5.1. INTRODUCTION: EXPLORING SECULAR MOVEMENTS

Kondratiev's (1925) hypothesis of the existence of long waves (with an average length of 50 years) have given rise to two distinct lines of historical research (Rosenberg and Frischtak 1984). One centred around price (or interest rate) cycles, and the other focused on long waves as a phenomenon in real quantities. It is the latter type of wave that this chapter will explore further. Long waves as a phenomenon in real quantities were sought by Kondratiev especially in relation to processes of capital accumulation (or investment), whereas others more in a Schumpeterian vein have sought for the role of technological innovation (or innovation clusters) with respect to economic 'booms' or crises. In various Schumpeterian interpretations of Kondratiev, waves of innovation clusters have mainly been studied in relation to two problems: Do they exist?[1] Which factors explain their occurrence and/or recurrence?[2].

However, in the 'long wave' debate in the Schumpeterian tradition, concerning the existence of innovation clusters, there has often been a puzzle in 'comparative studies' concerning the clustering of: (i) 'basic' (or radical) innovations (discussion taken up by Mensch, 1975); (ii) innovation flows (or fluctuations) more generally; and (iii) solutions of techno-economic problems around trajectories (using the terminology of Dosi, 1982) of innovation activity (discussion taken up by Freeman, Clark and Soete, 1982). This problem has not become less significant in related comparative studies asking why waves exist. Yet, such comparative studies (comprising both (i), (ii) and (iii) above) almost always have been in relation to (favouring or disfavouring) the approach of clustering of the *appearance* of 'basic' or radical innovations (i.e. approach (i)). This is especially true with respect to quantitative studies, where more or less adequate or satisfactory data on 'basic' or radical innovations have been the data source forming the arguments.

This chapter instead focuses on Freeman, Clark and Soete (1982) who argued how it is not the *appearance* of 'basic' (unrelated) innovations so much as their *diffusion*, which creates the clustering effect of related product, process, and applications innovations. Such views have also been related to Schmooklerian bandwagon effects (Schmookler, 1966), or associated with Schumpeterian 'entrepreneurial swarming' processes (Schumpeter, 1939). Freeman and Perez (1988) relate innovation clusters (c.f. long-term fluctuations) with the emergence of techno-economic paradigms.[3]

Hence, the first aim of this chapter is concerned with the third type of innovation wave debate mentioned above, i.e. the one related to solving techno-economic problems around trajectories of innovation activity. Based upon the technological trajectories derived from accumulated patent stocks in

---

[1] Kalecki 1939, Kuznets 1940, Kondratiev 1979, Forrester 1997, 1979, Kleinknecht 1981, 1987, 1990, Solomou 1986, Van Duijn 1981, 1983. (For a brief discussion, see section 5.2)

[2] Merton 1935, Schumpeter 1939, Schmookler 1966, Mensch, 1975, Freeman, Clark and Soete, 1982, Rosenberg and Frischtak, 1984. (For a brief discussion, see Section 5.2)

[3] Clearly, the concept of a 'techno-economic paradigm' or regime is much wider than that of 'clusters' (as defined by Freeman and Perez, 1988). See also Chapter 3 (section 3.2.1) for a discussion of the instituted nature of technological development governed by technological regimes or paradigms.

the previous Chapter 4,[4] it will be determined quantitatively whether there is evidence for periodic clustering of established innovation trajectories (rather than of 'basic' or radical innovations). Hence, we measure technological activity in certain technological areas, rather than specific innovations. Patent data are especially appropriate for such kinds of measures, since they are classified into schemes at a detailed level of aggregation, and over a long period of time. The type and nature of technological activity obviously varies over time.

This study, based upon establishing waves of takeoffs of innovation trajectories (derived from accumulated stocks of patents), also has some technical advantages in comparison to studies of 'basic' (or radical) innovations, in relation to what the innovation population consists of.[5] First of all, if an innovation trajectory within a technological field is finally formed and taken off, the innovations within it cannot be regarded as being arbitrary as a researcher's selection (or judgement of impact) of a specific 'basic' or radical invention or innovation.[6] It is argued here that, *ex post*, the successfully established innovation trajectories reflect both the areas and technological opportunities which possessed some degree of importance for the socio-economy. Secondly, established and identified innovation trajectories can *ex-post* be compared in size (impact), growth rate (opportunity) and life span (i.e. duration), so we can easily formulate some weighting procedure of the relative scope of each trajectory. Such information even allows us to gain insight into the structure of each wave of innovation trajectories, and a link

[4] By use of computational statistics logistic growth functions were fitted in Chapter 4 to US patent stock at a detailed level of disaggregation, including all Chemical, Electrical/electronic, Mechanical, Transport and Non-industrial technologies. Sustained periods of growth of patent stocks between 1920 and 1990 were traced and each trajectory was characterised in relation to the parameters which guided its evolutionary path. The results also justified the use of the now almost always assumed *S*-shaped image of the way technology develops within innovation trajectories, as well as confirmed the notion of the variety of technology dynamics.

[5] Kleinknecht (1990) stressed a serious problem in historical innovation research: 'In collecting "major" innovations (on whatever detailed definition) there is indeed a problem of how to deal with innovations of different (economic and/or technological) importance and complexity. The rate of innovations observed may be biased by the personal whims and preferences of the compiler' This reinforces Solomou's (1986) critique on the nature of the data on 'basic' or radical innovations, in the sense that some innovations are more basic than others, and we need a weighting procedure as the selection of such appears arbitrary.

[6] Also, as stated in Chapter 2 (section 2.2), it is believed here that by working with large numbers (trajectories of accumulated patent stocks in this case), the relative importance within the whole population tends to follow a normal distribution.

can be introduced between cyclical theories of technological evolution and the theories of innovation systems.

Hence, in this chapter, waves of takeoffs in innovation trajectories of Chemical, Electrical/electronic, Mechanical, Transport and Non-industrial technologies - as estimated in the preceding Chapter 4 - will be analysed in the context of overall innovation system takeoff. (The theoretical, conceptual and technical discussion concerning how the innovation trajectories forming the basis for this chapter are derived, will not be repeated in this chapter, and the reader should consult Chapter 4 for such information.) It is argued here that *ex post* the successful innovation trajectories taking off within a regime or paradigm reflect the areas of technological opportunities in which society possessed socio-economic competence, as without that the system would not have been unleashed.

In this respect, by only considering already established trajectories in which society has possessed socio-economic competence, this chapter only considers waves of takeoffs associated with socio-economic diffusion at the macro level. It has been suggested that foundations of such waves arise considerably earlier (Freeman and Perez 1988, von Tunzelman 1995).

The second aim of this chapter is 'descriptive' in nature, attempting to identify quantitatively as well as qualitatively historical waves of takeoffs in innovation trajectories. By use of panel-data analysis applied to the technological innovation trajectories estimated in Chapter 4, we are, in this chapter, able explicitly to compare timing of takeoffs and to rank the relative opportunity and impact of individual innovation trajectories within different epochs of trajectory takeoffs. From this evidence, the timing of unleashing of different innovation systems of technological trajectories, and their scopes (i.e. structure of opportunities and impact) will be identified. The main carriers and other technologies participating in the unleashing of the systems will also be identified. In this context, the overall configuration of an innovation system of technological trajectories can be defined as the boundaries of the main carriers and other technologies participating in its development, including their scopes (structure of opportunities and impact).

Some discussion on the timing of innovation system takeoff will obviously be incorporated when dealing with the two aims or problems of the chapter (i.e. do waves of innovation clusters exist? and if so what did they look like?), while the discussion concerning 'why' waves occur will be addressed more briefly.

The term 'innovation system' ought to be understood in a very *neutral* way, as an 'innovation system of technological trajectories'. The innovation system concept here is very different from the one used in the evolutionary economics literature on innovation systems, which is mainly based upon national innovation institutionalism (Freeman 1987, Nelson 1992) or national specificities in 'collective entrepreneurship' (Lundvall 1992). The systems of innovation in technological trajectories identified in this chapter will be linked in the next chapter to the notion of 'technological systems' developed by Carlsson and colleagues (1995, 1997).

Moreover, whereas Chapter 3 was concerned with the form and nature in which innovation opportunities changed across technological epochs governed by technological regimes or paradigms, this chapter is concerned with identifying and mapping the overall configuration of the innovation systems of technological trajectories, governed by such. As in Chapter 3, this chapter operates within the framework of *meta*-paradigms in the Freeman and Perez (1988) sense.

5.1.1. Chapter Outline

Firstly the data source or innovation trajectories used for the purpose of this analysis are presented. This will be developed from Chapter 4, which forms the empirical basis for this chapter. We then move on in the next section to discuss the possible empirical evidence for 'waves of innovation clusters' as provided by innovation trajectories of accumulated patent stocks. The notion of 'waves of innovation clusters' will then be conceptualised around a general unleashing of an 'innovation system of technological trajectories'. This section also presents how the structure (i.e. scope and overall configuration) of such a system might be measured quantitatively, by use of panel data analysis. Investigating the takeoffs in Chemical, Electrical/electronic, Mechanical and Transport innovation trajectories, the subsequent Section identifies and maps empirically the historical evidence concerning what such systems have looked like since 1920. Finally, some concluding remarks follow in the last section.

5.1.2. Data Selection

This chapter shares the Schumpeterian view that technology develops not along a smooth line but in cycles (Schumpeter 1939). Hence, the quantitative analysis in this chapter is appropriately based upon the 106 technological trajectories identified in Chapter 4. They are based upon time series of accumulated patent stocks 1920-1990 at the 56 technological group level (as identified in Chapter 2). However, for the purpose of this analysis, only Chemical, Electrical/electronic, Mechanical and Transport trajectories are used, and the trajectories belonging to the broad Non-industrial technological group are left out. This leaves us with 97 trajectories (See Tables 4.2-4.5, Chapter 4). The reason is that the Non-industrial technological group only has one technological category with sufficient impact and growth - i.e. nuclear energy - and this is considered separately in the discussion below. However, in addition to the 97 eligible trajectories, the more recent rising (still potential) trajectory of 'Pharmaceuticals and biotechnology' within Chemicals will also be included in the below analysis due to its great impact in recent research and developments, although this trajectory is not yet sufficiently formed.

5.2. WAVES OF INNOVATION CLUSTERS REVIVED

As explained in the Introduction (section 5.1), an investigation of the possibility of wave-like patterns in technological development is appropriate from a systemic perspective, especially in the context of overall innovation system takeoff (section 5.3. for conceptual framework). However, it is meaningful to first observe some rough and ready empirical evidence concerning clusters of innovation trajectories, as well as to discuss those in relation to the relevant theoretical literature within this field. This will provide a better starting point for both our theoretical and empirical understanding when analysing and discussing the data in a systemic context.

Table 5.1 Historical takeoff periods in innovation trajectories

Broad technological group	Early interwar	Postwar	Emerging
Chemical	1926 – 1932	1957 – 1970	1992 –
Electrical/electronics	1921 – 1934	1947 – 1977	1985 –
Mechanical	1922 – 1930	1950 – 1972	1985 –
Transport	1920 – 1930	1963 – 1978	-
TOTAL	1920 – 1934	1947 – 1977	1985

As mentioned in Chapter 4 (which provide the innovation trajectories informing this chapter), synchronisation of innovation growth cycles has often been used in tracing possible clusters of innovation. As we identified in chapter 4 a great diversity in the time span of each cycle, the cyclical phases are not appropriate to focus on when comparing the timing of innovation cycles. What is important is the point of technological takeoff. (Synchronised periods of takeoffs are subsequently (in section 5.3) associated with innovation system unleash.) We use the method of identification of innovation trajectories (or *S*-shaped technological growth cycles) and takeoff points, using US patent stocks as the measure of the cumulative stage of development, as described and identified in Chapter 4 (see Chapter 4 for technical measurement issues).

The takeoff points in trajectories of technological innovation, as identified in Chapter 4 (Graph 4.45), document two clustering periods in the twentieth century.[7] A closer look at the data might also suggest a third wave of takeoffs emerging, as seen in Table 5.1

It is striking that for long periods, like 1932-1957 and 1970-92 in the case of Chemical technologies, a time span of 25 years and 22 years respectively, there were no takeoffs. It is especially interesting to see how Transport trajectory takeoffs seem to be clustered, with no single takeoffs between 1930 and 1963, a period of 33 years. However, the patent data informing this analysis, ending in 1990, do not allow us to examine any takeoffs within Transport technologies since 1978 (see Chapter 4 (section 4.3.3) for model-fitting constraints), but it is evident that a substantial period with no takeoffs is experienced. Finally, whereas Mechanical trajectories also have longer periods with no takeoffs, of 20 and 13 years respectively, Electrical/electronics seems to have periods of takeoffs spread over longer historical periods, so that periods with no single takeoffs only cover 13 and 8 years respectively.

Hence, in total we have no innovation trajectory takeoffs between 1934 up to 1947, as well as between 1977 and 1985. The first period includes both the Great Depression and the Second World War. There could be many explanations for this.

For instance, a point which is commonly argued by historians is that takeoffs in technological innovations were held up by the Second World War and the (oil) crises of the 1970s. This would support an argument that takeoffs were otherwise randomly distributed. However, although external factors may account for much in innovative and economic fluctuations, this

---

[7] *Recalling Graph 4.45.* (Chapter 4) concerning the timing of technological cycle (or innovation trajectory) takeoff [technological group; year]*

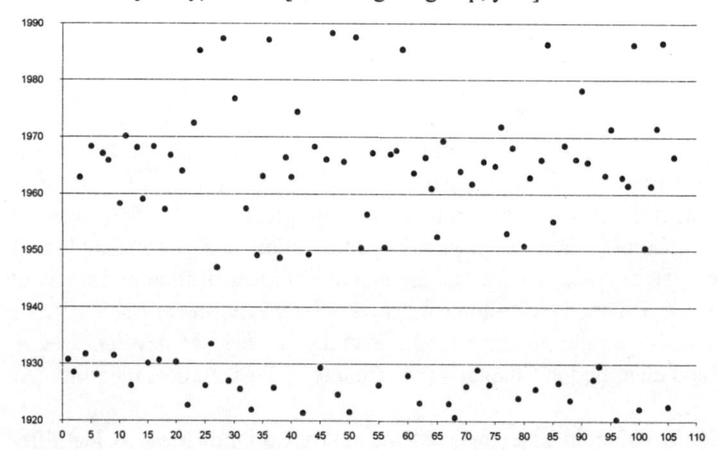

* *Technological groups are randomly numbered on the x-axis (i.e. they do not reflect the technological codes in Tables 4.3-4.7): Chemicals (1-21), Electrical/electronics (22-44), Mechanical (56-89), Transport (90-97) and Non-industrial (98-106).*

still leaves us without an understanding of whether and how the system by its own working produces clusters of innovation and booms or crises (issues addressed by Kuznets (1930a, 1940) and Schumpeter (1939)). Consequently (as argued in the introduction to this chapter) the most prominent discussions within evolutionary economics concern the possibility of clusterings of innovations (if not merely a statistical observation due to random historical events) being a by-product of the internal dynamics of the economic system. There are several potential strings to the latter argument.

If the clustering of trajectory takeoffs is treated as an indicator of clusters of innovations, it allows us to offer additional insights into the possible existence of long Kondratiev waves (Kondratiev 1925). That is, one could well argue that there was a Kondratiev upswing from the 1890s up to the late 1920s (there are only a few dots in Graph 4.45 in the period afterwards) interrupted only by the First World War, then a downswing of the third Kondratiev in the late interwar – Second World War period (i.e. late 1930s - early 1940s), followed by an upswing of the fourth Kondratiev in the 1950s and 1960s, and finally a downswing in the 1970s and 1980s. Such an argument would also be in line with the argument of Freeman, Clark and Soete (1982) and Kleinknecht (1990). However, others could argue that this upswing ended in 1913, and subsequently interpret the boom of the 1920s in terms of a reconstruction boom. Mensch (1975) suggested that depression-induced clusters of 'basic' innovations acted as the driving force for the next upswing, but his notion of 'basic' innovations is not really applicable in this context, as we are dealing with clusters of takeoffs in innovation trajectories.

Freeman, Clark and Soete (1982) argued that upswings are not to be related to the *appearance* of 'basic' innovations so much as to their *diffusion*, and that there would be a clustering of related product, process, and applications innovations making up the new technology and associated with the 'entrepreneurial swarming' process (rather than a clustering of unrelated innovations). This is supported by Coombs *et al.* (1987), who also highlight the fact that technological acceleration and diffusion in an upswing are very much related to institutional factors (i.e. no war, no financial crises, stable business environment etc.) rather than the extent to which the technologies are new or old ('basic' or not). In this perspective, innovation clusters are driven by a series of new socio-economic innovation systems, despite a more random distribution of 'basic' innovations. It can best be compared with clusters of technological trajectories using Dosi's (1982) terminology, rather than ('basic') radical inventions, in which case this study neither contradicts nor supports Mensch's argument.[8]

---

[8] In order to investigate whether there is a clustering of radical innovations in the downswing (Mensch's argument), it would be necessary to investigate in detail the

However, the evidence provided here, with respect to takeoffs in innovation clusters of patent stocks, seems to suggest that such takeoffs are well into Kondratiev's waves of economic upswing periods. Hence, it might be argued that the clusters in Graph 4.45 (Chapter 4) and in Table 5.1 above are demand-led 'Schmooklerian' or 'bandwagon' patent takeoffs related to investment demand in economic upswing periods.

From the literature we have also learned that innovation clusters may be a sectoral phenomenon with macroeconomic manifestations (Schmookler 1966, Schumpeter 1939), which derive their character from innovations in the industry where they begin; or, alternatively, they may be a 'real' phenomenon of long waves, associated with related induced innovations that gives rise to expansionary effects in the economy as a whole, in which clustering of innovations cannot be linked to a particular type of innovation as against other types carried out during the same epoch (c.f. Kondratiev's (1925) and Rosenberg and Frischtak's (1984) definition of 'long waves'). Without having identified the nature and structure of our innovation systems of technological trajectories it is hard to comment on any such argument at this stage.

The empirical evidence on innovation clusters presented in this section, will now be analysed in more detail and in a systemic context, providing insight into the possibility of waves as well as insight into the dynamics of the story.

5.3. THEORETICAL FRAMEWORK: UNDERSTANDING WAVES AS AN INNOVATION SYSTEM OF TECHNOLOGICAL TRAJECTORIES

When exploring wave like patterns in innovation trajectories, a system perspective can help us to derive the changing populations of technologies participating in the waves, and their relative structures; see theoretical framework below. Taking the changing population itself as the unit of analysis, and treating it as part of a wider system of relationships (c.f. the most essential function of a system perspective, see Chapter 1, section 1.3) also provides insight into the dynamics of the processes of system-development and structural change. In this way we will understand better if internally generated waves exist and what they looked like.

The theoretical framework of an innovation system of technological trajectories, and how it can be measured, will now be presented. It is in line

---

content of every patent within every technological field to identify which are the patents representing radical innovation. This is outside the scope of this chapter.

with the approach suggested in Andersen (1999b) and Andersen and Walsh (2000).

5.3.1. The Innovation System of Technological Trajectories and How it can be Measured

The technological frontier and the boundaries of an innovation system of technological trajectories.

An innovation system of technological trajectories does not have clearly defined boundaries, but, in a strictly technological sense, when there is a collection of technological innovation trajectories with an interrelated collective scope interacting at the 'technological frontier', then they serve as the base for the technological boundaries in the measurement of the system.

- In this context the technological frontier is measured as the point of takeoff on the *S*-shaped innovation trajectories (identified in Chapter 4).
- The boundaries of the innovation system of technological trajectories are measured by innovation trajectories which can be justified as being related, and taking off, within the same period of development.

Since broad periods of trajectory takeoffs are associated with broad frontier takeoff, they are in the following comparable with innovation system unleashes (given that they evolve with an interrelated and collective scope).

The scope of an innovation system of technological trajectories.

Not only are the technological boundaries a part of the configuration of the innovation system but so is the scope of each of the individual technological trajectories co-evolving within it. The scope of an innovation system of technological trajectories is related to the opportunity and the core importance of the technological trajectories within it.

- The 'opportunity' offered by a technological trajectory is measured by the percentage rate of growth in accumulated patent stock over the trajectory or growth cycle: a high opportunity is associated with a high rate of growth.
- The 'core importance' of a technological trajectory at the technological frontier is measured by the accumulated patent stock (c.f. accumulated socio-economic competence or historical importance) at the point of takeoff: a high stock is associated with a high degree of importance.

The overall configuration of an innovation system of technological trajectories.

The configuration of an innovation system is then measured by a collection of interrelated technological trajectories within a system (i.e. the boundaries), including the structure and composition of their opportunities and core importance or historical impact. Hence, trajectories with different time-spans, different opportunities and different impacts may participate in the unleash of a collective system, but each play a very different part.

It is the configuration of the innovation system of technological trajectories which defines the technological landscape for further innovation, and which defines the basic procedures concerning how best to exploit business opportunities. The configuration of the three periods listed in section 5.2, and which in Section 5.3 we call the launching of three innovation system epochs, are presented in Tables 5.2-5.4.

Main carriers and system participants.

The 'main carriers' of an innovation system of technological trajectories are here identified as those trajectories with the highest scope within the system. As the system scope is defined in relation to the structure of the trajectories' opportunities (trajectory growth rate) and socio-economic impact (accumulated patent stock at trajectory takeoff), Tables 5.2-5.4 highlight in bold the three trajectories within each broad technological group with the highest system opportunities as well as highest socio-economic impact at takeoff. However, due to small number problems, this does not apply to the Transport group, where only one trajectory was highlighted in the interwar period (only three trajectories were formed) and only two in the postwar period (only five trajectories were formed). Other technological innovation trajectories which were formed, and which also participated in the unleashing of the systems, are of course also important when understanding the full or entire scope of the systems, but as they are not defined as experiencing relatively high opportunities, or being of major socio-economic impact around takeoff, they are not here regarded as 'main carriers' of the systems and will be given less attention in what follows. Such trajectories can be termed 'system participants'. A reason for mainly focusing on trajectories defined as 'main carriers' in the following is an attempt at 'overview' (the line has to be drawn somewhere).

Considerations

Since there is assumed to be no single system movement at the frontier, but different movements of different technological trajectories at different rates and socio-economic impact, such a configuration of an innovation system implies an uneven type of system development, in terms of (i) the timing of takeoffs (although they seem to follow a certain band- (or epoch-like character), (ii) the periodicity of the time-span of each cycle (i.e. cycle duration), (iii) differences in opportunities of different technological fields, as well as (iv) cross-technology differences in socio-economic collective capability or historical impact. This may suggest that the formation of broader systems might evolve in an uneven fashion, in which dynamic forces within the formation of broader technological systems are created by internal tensions in which economics of scope can be achieved from accumulation, catching up and combinations of different technologies. This is also related to the system development phase and the solving of critical problems in this process (Hughes 1992, Dosi *et al.* 1988). Such interpretation is essential for understanding the dynamic elements of not just system development but also change and emergence (See Chapter 1, section 1.3). However, more empirical and theoretical elements concerning these very important issues are left out here, but instead carried out in the next chapter (section 6.3.), where the 'innovation systems of technological trajectories' identified in this chapter are analysed in the context of the notion of 'technological system dynamics'.

Table 5.2. Band of takeoffs in innovation trajectories: WAVE 1

Code*	Technological group	Takeoff year	Cycle duration (years)	Socio-economic competence at takeoff (Patent stock ranking)**		Cycle opportunity (Growth rate ranking)***	
CHEMICAL (1926–1932)							
2	**Distillation processes**	1931	98	456.01	(6)	**1727.46%**	**(1)**
3a	**Inorganic chemicals**	1928	15	**3157.18**	**(3)**	64.62%	(7)
4a	Agricultural chemicals	1932	6	350.96	(8)	46.17%	(8)
5a	**Chemical processes**	1928	23	**5901.40**	**(1)**	143.23%	(6)
7a	**Cleaning agents and other compositions**	1931	31	**4873.04**	**(2)**	176.91%	(4)
8a	Disinfecting and preserving	1926	35	54.72	(9)	161.81%	(5)
10a	**Bleaching and dyeing**	1930	20	1095.25	(5)	**229.57%**	**(3)**
51a	**Coal and petroleum products**	1931	25	2668.94	(4)	**562.32%**	**(2)**
55a	Explosive compositions and charges	1930	5	426.90	(7)	16.00%	(9)

*, **, ***: *See end of table*

Table 5.2. Continued.

Code*	Technological group	Takeoff year	Cycle duration (years)	Socio-economic competence at takeoff (Patent stock ranking)**		Cycle opportunity (Growth rate ranking)***	
ELECTRICAL / ELECTRONICS (1921–1934)							
30a	Mechanical calculators and typewriters	1923	6	4773.59	(6)	5.97%	(7)
33	**Telecommunications**	1926	22	6151.88	(4)	**86.71%**	**(2)**
34	Other electrical communication systems	1934	18	2276.23	(7)	39.33%	(5)
36a	Image and sound equipment	1927	9	5841.90	(5)	30.67%	(6)
37a	**Illumination devices**	1926	24	**8261.84**	**(3)**	**72.17%**	**(3)**
38a	**Electrical devices and systems**	1922	38	**13366.20**	**(2)**	**288.84%**	**(1)**
39a	**Other general electrical equipment**	1926	21	**14893.90**	**(1)**	69.95%	(4)
52a	Photographic equipment	1921	9	2044.61	(8)	4.90%	(8)
MECHANICAL (1922–1930)							
la	Food and tobacco products	1929	23	3769.34	(11)	73.06%	(4)
13a	**Metallurgical processes**	1925	28	7197.26	(8)	**126.93%**	**(3)**
14a	**Miscellaneous metal products**	1922	22	**79419.10**	**(1)**	19.74%	(12)
16a	Chemical and allied equipment	1926	17	26720.80	(5)	28.76%	(7)
18a	Paper making apparatus	1928	15	6961.61	(9)	50.65%	(6)
19a	Building material processing equipment	1923	11	4375.62	(10)	18.39%	(13)
22a	Other construction and excavating equipment	1923	4	2574.21	(14)	9.59%	(14)
22b	Other construction and excavating equipment	1930	6	2766.11	(12)	7.29%	(15)
23a	**Mining equipment**	1926	23	2603.99	(13)	**150.10%**	**(2)**
24a	**Electrical lamp manufacturing**	1929	16	126.07	(15)	**228.53%**	**(1)**
25	Textile and clothing machinery	1926	17	18788.80	(6)	19.79%	(11)
28a	Other specialised machinery	1924	16	33904.60	(4)	23.70%	(9)
29a	**Other general industrial equipment**	1926	15	**60275.90**	**(2)**	20.26%	(10)
50a	Non-metailic mineral products	1927	21	13362.80	(7)	57.79%	(5)
53a	**Other instruments and controls**	1924	18	**37063.20**	**(3)**	26.30%	(8)

* ** *** *See end of table*

Table 5.2. Continued.

Code*	Technological group	Takeoff year	Cycle duration (years)	Socio-economic competence at takeoff (Patent stock ranking)**		Cycle opportunity (Growth rate ranking)***	
TRANSPORT (1920–1930)							
44	**Aircraft**	1927	15	3240.01	(3)	**92.01%)**	**(1)**
45a	Ships and marine propulsion	1930	3	3777.05	(2)	6.08%	(3)
49a	**Rubber and plastic products**	1920	5	**7376.53**	**(1)**	16.52%	(2)

Main carriers of system participants are highlighted in bold.

* a, b, and c, after the technological group code number, denote different cycles of same technological group across different periods in time (see 'Takeoff year' columns).

** Socio-economic capability or importance at takeoff is defined as accumulated patent stock at the time of takeoff. See Chapter 4.

*** Cycle opportunity is expressed as the percentage growth in patent stock across the estimated cycle duration. See chapter 4.

Table 5.3. Band of takeoffs in innovation trajectories: WAVE 2

Code*	Technological group	Takeoff year	Cycle duration (years)	Socio-economic competence at takeoff (Patent stock ranking)**		Cycle opportunity (Growth rate ranking)***	
CHEMICAL (1957–1970)							
3b	Inorganic chemicals	1963	28	5062.29	(7)	134.10%	(7)
4b	**Agricultural chemicals**	1968	46	1511.31	(10)	**648.06%**	**(3)**
5b	**Chemical processes**	1967	35	**17701.50**	**(1)**	223.38%	(6)
6	**Photographic chemistry**	1966	61	4038.83	(8)	**1178.71%**	**(2)**
7b	Cleaning agents and other compositions	1958	12	9301.22	(4)	16.08%	(11)
7c	**Cleaning agents and other compositions**	1970	23	**11684.10**	**(2)**	62.44%	(9)
8b	Disinfecting and preserving	1968	42	212.85	(12)	333.49%	(4)
9	**Synthetic resins and fibres**	1959	64	**11488.70**	**(3)**	**1445.33%**	**(1)**
10b	Bleaching and dyeing	1968	17	2135.39	(9)	96.48%	(8)
51b	Coal and petroleum products	1957	3	7881.22	(6)	11.39%	(12)
51c	Coal and petroleum products	1967	12	8937.13	(5)	19.81%	(10)
55b	Explosive compositions and charges	1964	23	635.23	(11)	238.77%	(5)

*, **, ***: *See footnote to Table 5.2*

Table 5.3 Continued.

Code*	Technological group	Takeoff year	Cycle duration (years)	Socio-economic competence at takeoff (Patent stock ranking)**		Cycle opportunity (Growth rate ranking)***	
ELECTRICAL (1947–1977)							
30b	Mechanical calculators and typewriters	1972	5	2605.83	(9)	5.88%	(12)
35a	**Special radio systems**	1947	60	1883.58	(12)	**1379.59%**	**(2)**
36b	Image and sound equipment	1977	49	10555.30	(7)	303.64%	(5)
37b	Illumination devices	1957	4	10977.00	(6)	11.27%	(8)
38b	**Electrical devices and systems**	1949	4	**28472.60**	**(2)**	7.11%	(10)
38c	**Electrical devices and systems**	1963	24	**36214.70**	**(1)**	76.27%	(7)
39b	Other general electrical equipment	1949	2	20900.40	(4)	6.29%	(11)
39c	**Other general electrical equipment**	1966	30	**27013.30**	**(3)**	83.71%	(6)
40	**Semiconductors**	1963	26	2379.97	(11)	**1647.92%**	**(1)**
41	**Office equipment and data processing systems**	1974	56	15191.90	(5)	**865.29%**	**(3)**
52b	Photographic equipment	1949	3	2592.96	(10)	8.19%	(9)
52c	Photographic equipment	1968	42	4760.59	(8)	429.36%	(4)
MECHANICAL (1950–1972)							
lb	Food and tobacco products	1966	24	6194.82	(13)	72.09%	(8)
13b	Metallurgical processes	1966	24	13291.30	(10)	127.31%	(4)
15a	Food. Drink and tobacco equipment	1950	2	2943.09	(21)	5.79%	(22)
15b	Food. Drink and tobacco equipment	1956	5	1956.07	(22)	0.22%	(26)
15c	Food. Drink and tobacco equipment	1967	11	3042.89	(20)	21.90%	(18)
16b	Chemical and allied equipment	1951	2	26944.70	(7)	2.10%	(24)
16c	Chemical and allied equipment	1967	20	30815.50	(4)	48.82%	(10)
17a	Metal working equipment	1968	15	27595.10	(6)	35.92%	(12)
18b	Paper making apparatus	1964	22	8823.71	(11)	63.08%	(9)
19b	Building material processing equipment	1966	14	3054.75	(19)	26.89%	(16)
20	Assembly and material handling equipment	1961	34	23515.10	(8)	96.18%	(6)
21a	Agricultural equipment	1952	17	5899.76	(14)	15.32%	(19)
21b	Agricultural equipment	1969	4	6733.28	(12)	8.84%	(21)

*, **, ***: *See footnote to Table 5.2*

Table 5.3 Continued.

Code*	Technological group	Takeoff year	Cycle duration (years)	Socio-economic competence at takeoff (Patent stock ranking)**		Cycle opportunity (Growth rate ranking)***	
22c	Other construction and excavating equipment	1964	19	1934.24	(23)	32.38%	(14)
23b	Mining equipment	1962	25	5288.75	(15)	118.52%	(5)
24b	Electrical lamp manufacturing	1966	20	435.12	(26)	96.13%	(7)
26a	Printing and publishing machinery	1965	2	5015.73	(17)	5.32%	(23)
26b	Printing and publishing machinery	1972	9	5245.50	(16)	9.07%	(20)
27a	Woodworking tools and machinery	1953	13	1013.69	(25)	33.46%	(13)
27b	Woodworking tools and machinery	1968	16	1325.75	(24)	23.03%	(17)
28b	Other specialised machinery	1951	3	29046.00	(5)	2.09%	(25)
28c	**Other specialised machinery**	1963	23	**31832.20**	**(3)**	40.22%	(11)
29b	Other general industrial equipment	1966	15	**51296.90**	**(2)**	27.11%	(15)
31	**Power plants**	1955	35	3632.17	(18)	**237.37%**	**(1)**
50b	**Non-metallic mineral products**	1969	34	18231.40	(9)	**150.85%**	**(3)**
53b	**Other instruments and controls**	1966	63	**56444.60**	**(1)**	**220.520%**	**(2)**
TRANSPORT (1963–1978)							
42	**Internal combustion engines**	1978	27	**7012.02**	**(1)**	**192.07%**	**(1)**
43	Motor vehicles	1966	37	4056.74	(4)	137.50%	(4)
45b	Ships and marine propulsion	1963	23	2936.47	(5)	145.29%	(3)
46	Railways and railway equipment	1971	7	4669.33	(3)	16.28%	(5)
49b	**Rubber and plastic products**	1963	35	**5616.16**	**(2)**	**155.50%**	**(2)**

* ** *** *See footnote to Table 5.2.*

Table 5.4. Emerging band of takeoffs in innovation trajectories: WAVE 3 (This table is necessarily speculative.)

Code*	Technological group	Takeoff year	Cycle duration (years)	Socio-economic competence at takeoff (Patent stock ranking)**	Cycle opportunity (Growth rate ranking)***
CHEMICALS (1992–)					
12	Pharmaceuticals and biotechnology	1992	71	45220.90 (?)	1665.22 % (?)
ELECTRICAL / ELECTRONICS (1985–)					
30c	Mechanical calculators and typewriters	1985	16	2907.04 (?)	39.16% (?)
35b	Special radio systems	1987	5	5780.75 (?)	4.85% (?)
38d	Electrical devices and systems	1987	6	51151.70 (?)	7.96% (?)
MECHANICAL (1985–)					
1c	Food and tobacco products	1988	2	9271.75 (?)	7.29% (?)
14b	Miscellaneous metal products	1988	6	66335.70 (?)	11.55% (?)
17b	Metal working equipment	1985	5	32173.20 (?)	2.36% (?)
29c	Other general industrial equipment	1986	8	61295.50 (?)	3.40% (?)
TRANSPORT (–)		–			

*, **, ***: *See footnote to Table 5.2.*

As trajectories participating in wave-like patterns of innovation system takeoffs operate with very different time spans, and therefore mature at different times, it is difficult to talk about a mature state of a wave (c.f. for example Rostow's (1960) 'drive to maturity', and other theories evolving in different stages towards maturity). However, at the micro level the individual trajectories go through different stages of depression, revival (takeoff), prosperity (maturity) and recession (also maturity). (See Chapter 4 on conceptual framework for identifying trajectory (or cycle) properties.)

Another issue that is important to note is that, because innovations in technological patent groups are not seen to be participating in formalised trajectories, or because trajectories show no takeoff or rapid growth during a particular period, this does not mean that no inventive activity is happening. This may be the period when 'basic' or radical inventions are being patented, and individual patents are very significant for future exploitation, but not important in numbers (i.e. the field as a whole has not reached the collective competence necessary for an upswing, or revival). Periods with no takeoffs can also represent the time of prosperity for the trajectories within a system. Furthermore, although the technological trajectories in our analysis follow a common form of development trajectory, an *S*-shaped growth path, a technological growth cycle in this chapter is not a life cycle of the kind described by Kuznets (1930) and others, because it is now commonly recognised that old technological fields rarely die out. Instead, they often survive in niches, and get renewed in new paradigms and systems (see discussion and evidence in Chapter 3).

5.4. EXPLORING THE HISTORY

The configuration of the interwar and postwar systems, defined by the epochs of trajectory takeoffs (Tables 5.2-5.3), will now be discussed empirically for each broad technological group (Chemical, Electrical/electronics, Mechanical and Transport). Then we can speculate about the scope of the third wave of innovation trajectory takeoffs (Table 5.4).

5.4.1. Chemical Innovation Systems of Technological Trajectories

The main carriers of the interwar and postwar Chemical innovation systems of technological trajectories are presented in Tables 5.5 and 5.6, and they are discussed below.

In the interwar period, when many new chemical compounds were being obtained from coal in particular, innovations related to 'Distillation processes' grew at a rapid rate and offered substantial opportunities for commercial exploitation. (Patenting in distillation became less significant in terms of technological opportunity, or rate of growth, once petroleum took over as a feedstock.) 'Distillation processes' are applied to a wide range of chemicals, including organic acids, industrial alcohols from fermentation, and coal products. Since coal was the main feedstock of the chemical industry during the interwar period of takeoffs, the patents and technological opportunities in 'Coal and petroleum products' were concerned largely with exploitation of coal-based products, and processes for obtaining them. Hence, it was expected that this would be an area of great opportunities.

The interwar period was also one of major growth in output in synthetic dyes and related products.[9] This is consistent with Walsh (1984), who also notes a major growth in dye patenting in the period 1925-35. Although several new dyes were discovered during this time, most of the patents within 'Bleaching

[9] The period of basic innovations, with a lower rate of patenting but each patent representing a more significant technological (and even scientific) advance, had taken place before the data collected in this study (Walsh 1984).

and dyeing' were for new manufacturing processes, or for new dyes with improved properties such as fastness and solubility, and new shades.

Table 5.5. The main carriers of the Chemical INTERWAR innovation system of technological trajectories

Trajectories of greatest system opportunities	Trajectories of greatest socio-economic impact at system unleashing
'Distillation processes'	'Chemical processes'
'Coal and petroleum products'	'Cleaning agents and other compositions'
'Bleaching and dyeing'[10]	'Inorganic chemicals'

Table 5.6. The main carriers of the Chemical POSTWAR innovation system of technological trajectories

Trajectories of greatest system opportunities	Trajectories of greatest socio-economic impact at system unleashing
'Synthetic resins and fibres'	'Synthetic resins and fibres'
'Photographic chemistry'	'Chemical processes'
'Agriculture Chemicals'	'Cleaning agents and other compositions'

Although 'Coal and petroleum products' and 'Bleaching and dyeing' had developed some of their socio-economic impact (accumulated patent stock) already by the takeoff of the Chemical interwar system, it was different trajectories which dominated at the time of the unleashing of the postwar system.

Indeed 'Chemical processes' have been of significant socio-economic impact in both the interwar and postwar innovation system takeoffs, although it was not among those trajectories with the highest new opportunities. Whereas in the interwar 'Chemical processes' seem to have been mainly applied in relation to 'Coal and petroleum products', in the postwar system 'Chemical processes' were mainly used in relation to the rise of the new opportunities within 'Synthetic resins and fibres', as well as 'Photographic chemistry' and 'Agricultural chemicals'. In both the interwar and postwar system, 'Chemical processes' (with an estimated cycle duration of 23 and 35 years respectively) have also been an important technological facilitator in relation to invention and

[10] The technology was almost exclusively dyestuffs technology. Bleaching had been an innovative technology in the previous century.

innovations within 'Rubber and plastic products' (see the section on the broad Transport group below).

Trajectories of 'Cleaning agents and other compositions' have also been of significant socio-economic impact in both the interwar and postwar Chemical system takeoffs, although they were not among those trajectories with relatively high new opportunities. Hence, already at the start of the twentieth century, cleaning and related compounds with respect to personal, domestic and industrial hygiene had important socio-economic impacts.

Furthermore, within the interwar Chemical system takeoff, 'Inorganic chemicals' technology had a big frontier impact. Andersen and Walsh (2000) show how the reagents used in all chemical syntheses include 'Inorganic chemicals'. 'Inorganic chemicals' were one of the feedstocks in the heyday of the random synthesis of chemical compounds. It has also been argued that inorganic chemistry especially fed into the development of polyamide materials used in electronic generics. That is, polyamide applications in electronics traced from patent co-classifications (among technological patent classes of polymer chemistry, coating, polymer processing, chemical engineering, etc.) also included inorganic chemistry (Grupp and Schmoch 1992).

As much 'basic' patenting within 'Synthetic resins and fibres' had already happened in the interwar period, the technological field had grown into a significant patent stock when a trajectory took off in the postwar system.[11] This trajectory is estimated to be not only among those technologies with the largest socio-economic impact at point of takeoff, but also with the largest opportunity within the Chemical system. The trajectory of 64 years is estimated to end in 2002. 'Synthetic resins and fibres' was a dominant technology of the Chemical industry in the postwar technological system, and indeed one of the most dominant technologies of the economically advanced country economies in general. Freeman, Clark and Soete (1982) have discussed the takeoff of synthetic materials as one which fuelled the economic upturn in the postwar

[11] It was especially German scientists who did the basic research on the processes of polymerisation (i.e. forming of large molecules from repeated additions of small basic molecules or small groups of carbon atoms), and explored how such molecules could be produced artificially in laboratories in order to give materials certain desired characteristics. Herman P. Staudinger in Germany achieved an understanding of the processes of polymerisation which has formed the growth of the synthetic polymer industry since the 1930s. Walsh (1984, p.223) argued that his breakthrough represented a true scientific revolution. In 1953 Staudinger was awarded the Nobel Prize in Chemistry. By using Staudinger's principle of polymerisation an ever-increasing range of materials was produced synthetically from organic chemicals (first derived from the coal-tar industry, and then from petrochemicals).

period, and as one which offered great opportunities for economic and business development during this period.

Patenting within 'Synthetic resins and fibres' grew rapidly during the postwar period due to two major factors related to an enormous growth in demand (Andersen and Walsh 2000). The first was the reduction in price as a result of the use of petroleum as a feedstock for its production, not just because petroleum was cheaper than coal, but also because the monomers (intermediates) from which the polymers were obtained were also easier and cheaper to get chemically from oil.[12] The second reason was because problems of quality and suitability for new and replacement end-uses had begun to be solved by the use of additives such as plasticisers, stabilisers, antioxidants, extenders and fillers.

Patenting within 'Photographic chemistry' was very fast growing (i.e. had major opportunities) in the postwar period of system takeoff, but it did not have a relatively large socio-economic experience at the frontier during the takeoff period. Film and photography grew rapidly in this period, and reprographics started to be significant. Edgerton (1988) examines the close link between the chemicals and the photographic industry.

The postwar period was also a period of takeoff for the 'Agricultural chemicals (including pesticides)' trajectory, due to major renewed opportunities. Starting in 1954 and taking off in 1968, this trajectory is expected to last until 2000.[13] Likewise, Achilladelis *et al.* (1987) have shown that the 1950s and 1960s was the period when the most commercially successful pesticides were launched. Demand from farmers in the richer countries and aid-agencies and government programmes in the poorer ones, for products that would kill crop-destroying and disease-carrying pests rose sharply. The side effects that began to be revealed in the 1960s, leading to a ban on DDT in 1972 in the US, for example, generated further inventive effort directed at finding more environmentally friendly alternatives.

---

[12] Petroleum did not take over as the major feedstock for the chemical industry, even in the United States, until after the Second World War, though the US began to use it in the 1930s. Europeans (with plentiful coal but no oil until North Sea Oil was discovered) changed over much later. From Table 5.3 it can be seen that there were two further upsurges in patenting in 'coal and petroleum products' during this second period of takeoff in innovation trajectories (this time mainly petroleum), although with smaller opportunities than in the previous interwar system.

[13] Within the technological trajectory of 'Agricultural chemicals (including pesticides)' the patent class of 'Chemistry: fertilizers / pesticides) has increased its technological size ranking position (or socio-economic impact position) to become among the top 10 highest ranked in 1990 among the total 399 patent classes (i.e. including all Chemical, Electrical/electronic, Mechanical and Transport and Non-industrial technologies).

5.4.2. Electrical/electronic Innovation Systems of Technological Trajectories

The main carriers of the interwar and postwar Electrical / electronics innovation systems of technological trajectories are presented in Tables 5.7 and 5.8, and they are discussed below.

Table 5.7. The main carriers of the Electrical/electronic INTERWAR innovation system of technological trajectories

Trajectories of greatest system opportunities	Trajectories of greatest socio-economic impact at system unleashing
'Electrical devices and systems'	'Other general electrical equipment'
'Telecommunication'	'Miscellaneous metal products'
'Illumination devices'	'Other instruments and controls'

Table 5.8. The main carriers of the Electrical/electronic POSTWAR innovation system of technological trajectories

Trajectories of greatest system opportunities	Trajectories of greatest socio-economic impact at system unleashing
'Semiconductors'	'Electrical devices and systems'
'Special radio systems'	'Electrical devices and systems'*
'Office equipment and data processing systems'	'Other general electrical equipment'

* *i.e. 'Electrical devices and systems' had two trajectories in the postwar system.*

Electrification and the discovery of the electron have probably been one of the most consequential technological changes over the last century. In the twentieth century electricity has put virtually unlimited power at the disposal of all, and with the discovery of the electron, the formation of electronic communication and information technologies were initiated.

What is indeed evident are the numerous electrical and electronic devices that the development and distribution of electricity brought about. 'Electrical devices and systems' was in the interwar system among the trajectories with both high socio-economic frontier impact (i.e. high accumulated patent stock) as well as an area with great new opportunities (c.f. high growth rate). In subsequent periods of system takeoffs this trajectory has been an area of core impact at the technological frontier. 'Other general equipment' was also an area of great frontier impact, rather than an area of new opportunities; this applies to both the interwar and postwar systems. Hence, these two technological groups contain a variety of technological feedstocks facilitating the unleashing of the Electrical/electronic innovation systems.

As the 'basic' innovations relating to the light bulb were already invented well before the end of the nineteenth century,[14] 'Illumination' technologies had already grown large in terms of socio-economic impact at the unleashing of the interwar innovation system. The trajectory of 'Illumination' first took off with the diffusion of electricity, in which it became a central area of new opportunities.[15] Within the Mechanical broad technological group (see Section 5.3.3. below) it is also interesting to see how 'Electrical lamp manufacturing' was the area of greatest new opportunities in the interwar system, although it was not a trajectory of very high socio-economic impact at the unleashing stage.

Although the 'Telecommunications' trajectory was not very high in socio-economic impact at the unleashing of the interwar system, it certainly appeared to be among those with the highest system opportunities. By investigating the structure of patenting within 'Telecommunications' during the interwar system, it is evident that it was not telegraphy and telephonic communications which grew with high rates, but rather another related branch - namely pulse or digital communications. Furthermore, the telephone's acoustic wave system became essential in the interwar period for inventions concerning practical systems of recording and reproducing sound, and within 'Telecommunications' especially patent classes of acoustic wave systems and related devices were fast growing.

Soon after development within telecommunication technologies (e.g. the telephone, including its acoustic wave systems) the idea of broadcasting

---

[14] The basic inventions of the light bulb were by Thomas Edison and Joseph Swan between the years 1878-80.

[15] However, in recent times a new type of lighting - Light Amplification by Stimulated Emission of Radiation (i.e. LASER technology) - has been of significant importance, and this technology has integrated with the development of a variety of rather different technological fields (especially optics). It is argued that in many cases it has been substituting for other light sources without changing the purpose of the light apparatus. Hence the utility of the inventive step is proven in terms of new integrated products and processes as well as in terms of better performance characteristics of the LASER product in comparison to existing products (Grupp and Schmoch, 1992). LASER technology (or coherent light generators) is patented within the technological group 'Electrical devices and systems', rather than the group of 'Illumination'. Patenting of coherent light generators within 'Electrical devices and systems' first started in the 1960s, but has since then grown to having major socio-economic impact (out of the 399 patent classes in total, across all Chemical, Electrical/electronic, Mechanical, Transport and Non-industrial patent classes, 'Coherent light generators' were in 1960 ranked 396 in terms of accumulated patent stock; by 1990 its technological size ranking had increased to 121).

messages was quick to follow. Hence, opportunities in postwar takeoffs of innovation trajectories were very high within 'Special radio systems.'[16] Especially the radio set, the television, RADAR (RAdio Detection And Ranging) and satellite telecommunications were central. That is, by investigating the patent classes within the trajectory of 'Special radio systems', those classes containing communication technologies, including directive radio wave systems and devices (e.g. RADAR; radio navigation), as well as radio wave antennaes, are those which have accounted for the high opportunities (or growth rate) of 'Special radio systems'.[17] It has been argued by Day (1990, p.219) that the chemical polymer industry especially made innovations in electrical and electronic engineering manageable - which would not otherwise have been possible - and that this in particular applies to innovations in the field of RADAR. From patent co-classifications Grupp and Schmoch (1992) also traced polyamide applications in electronics

Although many improvements have been made in radio broadcasting (including advances in detectors and transmitters, as well as antennae locating and retaining the signals from broadcasting stations)[18], it was not until 1948

---

[16] Although Heinrich Hertz had proved the existence of electromagnetic waves and wire-less transmission already in 1887 using what was properly the first antenna (hence the means to generate and detect electromagnetic waves had been found well before the twentieth century), the exploration of the Hertzian waves by many different scientists and corporations has been a twentieth-century phenomenon.

[17] At the beginning of the 1930s military applications of radio waves were investigated, pushing the development of RADAR technology in both the US and several European countries (especially Britain and France), and the first practical RADAR appeared in the UK in 1935. It was the start of a defence system based on electronics, and in the beginning it was in particular used for guidance of aircraft and as a means of detecting and locating enemy aircraft, but it was also used for locating U-boats and ships. However, RADAR technology has also become essential not only for military purposes, but also for civilian air traffic control, advanced radio communications, meteorology and new forms of science (radio astronomy).

[18] However, the 'basic' inventions for these technologies took place well before. The first patent for the transmission of wireless radio telegraphy (transmission of messages without wires) was granted in 1895 to Guglielmo Marconi, who established the Marconi Wireless Telegraphy Company in 1897, and few years after, in 1909, Marconi got the Nobel Prize in physics. The first documented broadcast of speech and music was several years later on Christmas Eve 1906 from Brant Rock, Massachusetts - the transmitter was built by General Electric Company. However, the first US radio licence specifically for broadcasting was given to a Westinghouse station in Pittsburgh in 1920 (Ohlman 1990, p.726). Chandler (1990, p.218) focuses on General Electric, Westinghouse and Western Electric as the pioneers in radio receivers and broadcasting equipment. The first frequency broadcasting in the US was allocated to Amplitude Modulation (AM) broadcasting; but in 1933 the Frequency Modulation

when the Bell Laboratories invented the transistor (after research on germanium and silicon) that the generics became compact and simple enough to make truly portable radios (i.e. transistor radios). A transistor is a small semiconductor, and it caused a revolution in electronics because of the complexity of circuits that could be packed into small spaces and its very low heat emission characteristics. Furthermore, it needs very little electricity and can run for hours on small batteries.

The technological development of semiconductor technology is also evident from Table 5.3, in which we see how the trajectory of 'Semiconductors' experiences the highest growth (or opportunities) in the postwar Electrical/electronic innovation system. The invention of the semiconductor was also central in relation to the development of 'Office equipment and data processing systems' which also formed an innovation trajectory of great opportunities in the postwar system. The semiconductor revolution here was mainly in relation to the development of the Integrated Circuit (IC) where thousands of tiny semiconductors or transistors are packed into a single chip of silicon, which culminated in the development of the desktop computer.[19] The ICs have since been the 'basic' building blocks of digital technology in general. Also, from a closer look at the patent data, we see how the process of manufacturing semiconductors enjoyed great opportunities, as the patent class of the semiconductor device manufacturing processes has been among those accounting for high growth within 'Other general electrical equipment' since 1940.

Hence, with the emergence of the micro-electronic revolution, the above mentioned office and business machines have become electro-mechanical, or purely electronic, in operation, and the overwhelming opportunities for 'Office equipment and data processing systems' (whose current Electrical/electronic innovation trajectory is estimated to end in 2013) has been strongly intertwined with the development of other very successful technological fields such as telecommunications and semiconductors. The trajectory of 'Office equipment and data processing systems' includes technological fields such as electrical computers and data processing systems, static and dynamic information storage or retrieval, and electronic funds transfer.

(FM) radio system was patented, and in 1938 General Electric installed the first FM broadcasting station (Ohlman 1990, p.728). (In contrast to AM, FM is noise-free and the preferred method for broadcasting stereophonic sound.)

[19] The first electronic computers were completed in the 1940s. Electronic computers were first based on thermionic valves, punched cards for programming and magnetic tapes for recording. They took up a lot of space and were completely dependent on the mainframe for operation. Rapid technological developments were taking place in order to provide independent computers in smaller packages such as desktops (Cortada 1993).

5.4.3. Mechanical Innovation Systems of Technological Trajectories

The main carriers of the interwar and postwar Mechanical innovation systems of technological trajectories are presented in Tables 5.9 and 5.10, and they are discussed below.

Table 5.9. The main carriers of the Mechanical INTERWAR innovation system of technological trajectories

Trajectories of greatest system opportunities	Trajectories of greatest socio-economic impact at system unleashing
'Electrical lamp manufacturing'	'Miscellaneous metal products'
'Mining equipment'	'Other general industrial equipment'
'Metallurgical processes'	'Other instruments and controls'

Table 5.10. The main carriers of the Mechanical POSTWAR innovation system of technological trajectories

Trajectories of greatest system opportunities	Trajectories of greatest socio-economic impact at system unleashing
'Power plants'	'Other instruments and controls'
'Other instruments and controls'	'Other general industrial equipment'
'Non-metallic mineral products'	'Other specialised machinery'

Concerning the broad Mechanical technological group, 'Other general industrial equipment' and 'Other instruments and controls' have been of high core impact, or socio-economic importance, at the unleashing of both the interwar and postwar Mechanical innovation system. This is not so surprising since those are all wide-ranging technological trajectories containing a variety of technological sectors, and must be expected to be of general importance for the unleashing of almost any broad Mechanical system. This is also true for 'Other specialised machinery' during the postwar takeoff period.

Furthermore, 'Other instruments and controls' is the technological trajectory in the postwar Mechanical innovation system with not only the largest technological frontier impact (i.e. patent stock at takeoff), but also the fastest growing (i.e. with the greatest opportunities). The trajectory is estimated to end in 2009. By viewing the data, the new opportunities within 'Other instruments and controls' in the postwar system are especially caused by inventions within areas such as recorders, optics (with respect to surgery, dentistry, eye examining; vision testing and correcting, communication systems, measuring and testing), horology (time measuring systems or devices); while the largest contribution to the patent stock (accumulated from the past) or socio-economic capability at takeoff include machine elements, technologies of measuring and testing (especially electricity) as well as geometrical instruments, surgery, and optics (with respect to communication systems and elements). This illustrates that the patents accumulated from the interwar period are mainly related to engineering, electrical and (tele)communication systems, whereas the opportunities in the postwar system are more related to new kinds of optics.

Furthermore, it is interesting how 'Metallurgical processes' and 'Miscellaneous metal products' in the interwar system come out as areas of high system opportunities and of great frontier importance or socio-economic impact, respectively. Also 'Mining equipment' seems to be an area of high importance at the unleashing of the interwar system takeoffs. The trajectory in 'Mining equipment' technologies might be related to the trajectory takeoff in 'Coal and petroleum products' within the broad Chemical technological group, as well as those related to metal technologies mentioned above. Table 6.1 (Chapter 6) also indicates how the broad industrial sectors of 'Coal and petroleum products' within the broad Chemical industrial group, as well as 'Metals' within the broad Mechanical industrial group, specialise within 'Mining equipment' in comparison to other industries.

With respect to the interwar Mechanical system, 'Electrical lamp manufacturing' seems to be an area of great opportunities. This seems to be naturally coexisting with both the core impact and opportunities of 'Illumination devices' within the broad Electrical/electronic technological group, which was mentioned in the section above (together with the Electrical/electronic innovation systems of technological trajectories).

As also described in the section above, at the unleashing of the postwar system, modern life indubitably came to depend on electricity (for industry, commerce, transport, street lighting, and the home - domestic electricity for space heating, water heating, cooking, freezing, refrigeration, lighting, and motorised appliances such as vacuum cleaners, washing machines, food mixers, etc.). As demand for electricity increased for industry and in cities and well as for transport infrastructures, power plants and powerful engines were needed to drive generators (Hughes 1983). By the postwar system takeoff, an innovation trajectory within 'Power plants', and with very low socio-economic capability at takeoff, experienced high innovation opportunities, and the innovation cycle lasted for 35 years until 1979. Although power plants were initially developed to be driven by steam turbines (the improved steam turbine made thermal power stations possible by the turn of the century (Du Boff 1967)), the power plants in the postwar system were mainly driven by coal or petroleum, as well as driven by nuclear reactors at a later stage.

Although the Non-industrial broad technological group does not enter into this chapter it is here relevant to mention how enthusiasm often has been aroused by the prospects of very cheap energy, and how nuclear energy was explored for utilisation as atomic power driving power engines. As can be observed in the data from Chapter 4 (Table 4.6), innovations within 'Nuclear reactors' (complementary to those within power plants), taking off in 1961, form a trajectory estimated to have lasted until 1974. It is the trajectory with the highest opportunities across all 106 trajectories developed throughout the last century.[20]

Finally, an innovation trajectory of 'Non-metallic mineral products' (mainly glass and advanced ceramics) lasting for 34 years also took off in the postwar system. This was among the trajectories with the greatest opportunities, and is estimated to have lasted until 1992. There are a number of underlying reasons for the great increase in patenting within 'Non-metallic mineral products' in the postwar period (Phillips, forthcoming). Fundamental to this was the introduction of float glass processes that resulted in both an increase in production and quality. Also, the automobile industry is inextricably linked to the glass industry, and the increased demand for glass surfaces in cars would have resulted in an increased demand for glass products. Glass production also continued to rise through continued demand for double-glazing. As well as this increased usage of glass, there has been a move towards the development of more specialised glass products. Stringent safety regulations have brought about a demand for toughened safety glasses and, in the

[20] The basic structure of the atom was identified by the Ernest Rutherford and Niels Bohr just before the First World War, and with them the most characteristic branch of the twentieth century physics - nuclear physics - was born. There were speculations on how the immense binding energies that hold the nucleus together could possibly be exploited as atomic energy or as new sources of power, and applying Einstein's special theory of relativity from 1905, the energy released by a so-called chain reaction can be enormous. Nevertheless, nuclear physics was still a 'pure science' until the end of the Second World War, when an interdisciplinary research team including J. Robert Oppenheimer, Niels Bohr, U. Enrico Fermi, James Chadwick and J. von Neumann found how enormously condensed nuclear energy could be utilised for practical purposes, such as explosive super-bombs leading to the development of atomic bombs or atomic power driving power engines. Turbines driven by steam raised by the heat of a nuclear reactor can be used to generate electricity. Just for clarification, patents within 'Nuclear reactors' do not cover nuclear material or atomic energy for atomic weapons. The Atomic Act of 1954 (United States Patent and Trademark Office) excludes patenting of inventions developed for that purpose.

construction industry, the call for energy efficient buildings has stimulated the development of low emission glasses (such as 'K'-glass produced by Pilkington and Glaverbel under licence). The development of the IT industry has also been interrelated with inventions and innovations within 'Non-metallic mineral products', with glass as an important component of all information technology products, like fibre optical networks, computer screens and magnetic glass disks.

5.4.4. Transport Innovation Systems of Technological Trajectories

The main carriers of the interwar and postwar Transport innovation systems of technological trajectories are presented in Tables 5.11 and 5.12, and they are discussed below.

Table 5.11. The main carriers of the Transport INTERWAR innovation system of technological trajectories*

Trajectories of greatest system opportunities	Trajectories of greatest socio-economic impact at system unleashing
'Aircraft'	'Rubber and plastic products'

* *Only one trajectory is mentioned here due to small number problems (see section 5.3).*

Table 5.12. The main carriers of the Transport POSTWAR innovation system of technological trajectories*

Trajectories of greatest system opportunities	Trajectories of greatest socio-economic impact at system unleashing
'Internal combustion engines'	'Internal combustion engines'
'Rubber and plastic products'	'Rubber and plastic products'

* *Only two trajectories are mentioned here due to small number problems (See section 5.3).*

Within the interwar innovation system, the 'Aircraft' innovation trajectory (which mainly includes aeronautics) illustrated particularly high inventive activity, or particularly high technological opportunities, being the fastest growing trajectory within transport at that time.[21] Likewise, Mowery and

[21] The basic inventions and innovations of the aeroplane were completed by the Wright brothers (Orville and Wilbur Wright) who in 1903 invented the first true

Rosenberg (1982, p.166) state that an increasing level and quality of research was done within aeronautics in the interwar period. It has even been argued that in the interwar years the aeroplane was seen as one of the most important products of modern science and technology, and one which would have particularly profound consequences for warfare, societies and economies (Edgerton 1991). However, in the interwar period the development of civil aviation was limited to the wealthy. Hence civil air transport could only exist if subsidised by military orders or maintained by airmail contracts (Mowery and Rosenberg 1982).

Furthermore, concerning interwar trajectories, within transport 'Rubber and plastic products' were certainly among those with high socio-economic frontier impact (accumulated from inventions from the past), although not of very high new opportunities.[22] Up to the interwar period, this technological trajectory was mainly related to the development of rubber for tyres for the rapidly emerging automobile industry. According to Chandler (1990, p.105), the demand for tyres increased a hundred-fold from 1910 to 1930.

However, in the postwar innovation system, 'Rubber and plastic products' were not only of relatively high frontier impact or importance, but this time also experienced high renewed opportunities for new developments. The inventions here were mainly in relation to plastics, rather than rubber, which were a mature technology at this stage. First plastics were used as replacement materials for ceramics, metal, glass or wood, then they were applied to new end-uses in order to exploit their unique advantages over other materials, such as electrical resistance, lightness, plasticity, and mouldability. However, the 'Rubber and plastics products' innovation trajectory was not significant in core impact or opportunity relative to 'Synthetic resins and fibres' within the broad Chemical technological group, because most of the inventive work was not in the plastics products themselves, but in the materials and processes for making and fabricating them, covered by the 'Synthetic resins and fibres' category. The patents in this area, in the 1950s and 1960s, were overall mainly concerned with processes, modified products, fabrication techniques and new applications (Andersen and Walsh 2000).

Trajectories within the Transport postwar system especially seemed to reflect the development of the motor-car industry. 'Internal combustion engines' were central in this system as the trajectory with both the largest socio-economic

---

aeroplane (with propellers and powered by a gasoline engine), and in 1905 they took their first passengers on a flight.

[22] The really inventive period for natural rubber was at the turn of the century, but basic research on polymers (e.g. by Herman P. Staudinger in the 1920s, see section 5.4.1) gave rise to renewed interest in finding a commercially viable synthetic rubber, and improved the possibility of doing so.

impact at takeoff, as well as with relative highest opportunities. That is, a number of successful engines which use different fuels such as water, wind, gas, gasoline or petrol, powdered coal or other solid fuel and oil have been developed over time. However, the steam turbine was gradually being challenged by the electric motor, but eventually the much more powerful internal combustion engines (which all relate to one family of engines working on the same principles) managed to out-compete the electric motor. The problem with the steam engine was that it was not compact enough (too big) for many application purposes outside power plants (e.g. application in motor vehicles, ships and aircraft), and the electric motor was (is) not yet powerful enough. But in terms of speed, power and light weight, the petrol or gasoline and other 'Internal combustion engines' ensured their ultimate success, although it took several years before the innovation trajectory was defined and took off. Within the postwar system its growth cycle is estimated to survive (or have survived) until 1997.

5.4.5. Speculations on an Emergent Overall Innovation System of Technological Trajectories

Due to data constraints it is difficult to conclude when the second epoch of trajectory takeoffs ended and when a new one began.[23] However, as stated in section 5.2, due to a 'fading out' or gap in the innovation trajectory takeoffs from the 1970s to the late 1980s, it is here suggested (or speculated) that a third innovation system of technological trajectories might have emerged in the mid/late 1980s.

In this emergent system of takeoffs in innovation trajectories, it is astonishing to see in Table 5.3 how the 'Pharmaceuticals and biotechnology' trajectory (which is included here although it was initially omitted due to data constraints – see footnote 23) has high opportunities similar to those of 'Distillation processes' in the interwar period, as well as 'Synthetic resins and fibres' and 'Semiconductors' in the postwar innovation system. However, in terms of socio-economic impact at the frontier takeoff, none of the latter system carriers (i.e. 'Distillation processes', 'Synthetic resins and fibres' or 'Semiconductors') can even compare with the large accumulated patent stock (or socio-economic competence) of 'Pharmaceuticals and biotechnology'. Starting in the early

[23] It ought to be mentioned that the few takeoffs in the late 1980s may be due to data constraints (the data source informing this chapter ending in 1990). That is, by definition the last empirical observation of the cycle has to be at least after 60% of the long-run estimated technological cycle potential in terms of accumulated patent stock (see Chapter 4, section 4.3.3). This includes even those technological groups whose cycles break empirically in 1990 owing to data constraints.

1970s and taking off in 1992, the trajectory of 'Pharmaceuticals and biotechnology' is estimated to last until the 2040s.

However, there has been some debate about whether biotechnology will be able to fuel a new economic upswing in the way that synthetic materials did (see for example Walsh 1993). Demand is obviously a key issue here. Biomass is a technologically viable feedstock for chemicals and the synthesis of fuel alcohol, but if it is to become commercially viable it might be because petroleum has become much more expensive, or because changes in environmental policy world-wide have severely restricted the use of petroleum (Andersen and Walsh 2000).

In any case, 'Pharmaceuticals and biotechnology' within the Chemical technological group is starting to offer a wide range of opportunities for sectors of the Chemical and allied products industry, such as pharmaceuticals, agrochemicals, waste treatment and foods. Biotechnology can offer new products (from genetically modified organisms to diagnostics), new processes (e.g. genetic engineering to make human proteins such as insulin, cell culture to make plant-based drugs and other products) and new product search methods for application in research and development. Biotechnology is not only a new technology for the Chemical industry, but a new area of expertise for its staff, representing a change from synthetic chemistry to the life sciences. Also, although patent stocks in food technology (i.e. 'Food and tobacco') have begun to take off during this period, both the rate of growth and the accumulated patent stock is much lower than in the category 'Pharmaceuticals and biotechnology'. However, many of the innovations in food technology are being made under the heading of biotechnology (and patented in this category) (Andersen and Walsh 2000).

Concerning the other trajectories identified in the emergent systems of Electrical/electronics, Mechanical and Transport, none of the technological trajectories are relative large in opportunity or cycle duration (and are therefore left out in this discussion).

It is obviously expected that the more longer-term trajectories or system main carriers are not yet identified (due to data constraints, as identified in footnote 23). Alternatively, and as likely, the structure of the next upswing is still under exploration (i.e. search and selection), and trajectories have not yet been formed, nor taken off.

5.5. CONCLUSION

Much of the quantitative academic literature on 'waves' of innovation clusters has focused almost exclusively on 'basic' or radical innovations; even when interpreting the notion of waves in line with Freeman *et al.* (1982), who even argue that the focus on their *appearance* has very little to say about why innovations clusters occur. Freeman is dealing with the processes throughout the socio-economic system, in the generation of innovation clusters through the formation and diffusion of pervasive trajectories.

We have in this chapter focused exclusively on the 'takeoffs' of related already established innovation trajectories, as opposed to 'basic' or radical innovations. How such takeoffs are associated with diffusion is confirmed in Chapter 8, which deals with the relationship between diffusion of corporate technological leadership in the course of technological growth.

Working with already formed established innovation trajectories, as opposed to 'basic' innovations', also helps us to overcome a crucial issue with respect to selecting and weighting the 'innovation population'. Hence, the results become less arbitrary.

Combining patent statistics in Chapter 4 with panel data analysis in this chapter, we have overall traced two waves of takeoffs in innovation systems of technological trajectories, and suggested the emergence of a third. The overall interwar innovation system gave scope to a system based on coal as a feedstock, chemical process technology and technology related to products such as rubber (especially for tyres) and dyes. Along with this Chemical system emerged electrical engineering, telecommunication, a narrow range of optics, and a broad range of illumination. Developments also took place within heavy metal engineering and mining, as well as in aircraft technology. This evidence drawn from patent records also supports Freeman and Perez, who have termed this period the 'Electrical and heavy engineering [third] Kondratieff' (Freeman and Perez 1988, p.51).[24]

The overall postwar innovation system was a much more complex system with extended interrelated technological boundaries at the broad technological group level.[25] The system was based on petrochemical feedstocks, synthetic materials, and plastics with broad technological applications. Non-metallic mineral products, such as glass and advanced ceramics, also gained great opportunities. All those developments were especially interrelated with radio-systems (especially RADAR technology), office equipment and data processing, as well as new optics. It was an energy heavy system with major growth in power plants (including nuclear technology). Development of engines and fuel was also central, especially for motor vehicles. Again here, the evidence drawn from patent records supports

---

[24] As statistically identified in Chapter 3, this period was characterised by intra-group technological diversification and the formation of a structure of specialised engineering and science-based fields, as opposed to more complex cross-technology integrated evolution.

[25] The statistical evidence for this is also confirmed in Chapter 3.

Freeman and Perez who have termed this period the 'Fordist mass production [fourth] Kondratieff' (Freeman and Perez 1988, p.52).

Finally, this study forecasts an emerging system which moves us into the life sciences and the exploitation of biotechnology. In a Kondratiev context, it has been suggested (or speculated) by Freeman and Perez (1988, p.53) that a fifth 'Information and communication Kondratieff' has started during the 1980s and 1990s. They suggest that micro-electronics, biotechnology products and processes, as well as fine chemicals are among the key factor inputs. So far it is still early days for telling whether the patent story in this chapter supports or contradicts this argument, but the patenting record of pharmaceuticals and biotechnology is evident. However, as suggested by Andersen and Howells (2000), the extent to which industrial patents are, or will be, applicable to use as main indicators in the context of much new information technology, such as software and related, is doubtful.

Accordingly, the waves of innovation takeoffs in technological trajectories are best compared with Freeman and Perez's (1988) interpretation of the third, fourth and fifth Kondratiev. The results also indicate that the timing of takeoffs in the technological trajectories tends to be clustered within periods normally associated with economic upswing (as also suggested by Freeman, Clark and Soete, 1982). Yet, the trajectory takeoffs seem to be well into the economic upswing periods.

This might suggest that the waves of trajectory takeoffs are best compared with Schmooklerian demand-led bandwagon effects, related to investment booms in economic upswing periods. However, the evidence presented is much too tenuous to be worth drawing any such definite conclusion concerning *why* the unleashing of innovation systems occur. Also, a validation of whether the documented waves of trajectory takeoffs were due to 'sectoral phenomena with macroeconomic manifestations' or real 'long waves in economic life' (in the Kondratiev (1925) sense), would require further investigation into the relationship between invention, innovation, economic multiplier effects and externalities, as well as into an interpretation of history. Such analysis is indeed interesting, but outside the scope of this chapter.

Although the evidence in this chapter supports Freeman, Clark and Soete (1982), that innovation trajectories cluster in upswing periods, associated with solving of techno-economic problems in support of system unleashing, it cannot be regarded as an extended criticism of Mensch's (1975) idea of clustering of depression-induced innovations, since we do not consider 'basic' or radical innovations. Nevertheless, in the sections (especially footnotes) concerning how the main carriers of the Chemical, Electrical/electronic, Mechanical and Transport innovation systems of technological trajectories looked like at their unleashing, we did learn how the 'basic' inventions often happened several decades before trajectory takeoff. This supports Kondratiev, who believed that important discoveries must wait for the next wave's upswing to be implemented on a large scale.

Although the empirical findings in this chapter are of course not new, it is the first time they are revealed quantitatively in a system framework, and based upon a century of patenting activity. Using an innovation system approach as the 'analytical tool' in addition to a 'descriptive framework' was especially useful in illustrating in more detail, and more accurately, how the scope (opportunities and core-importance) and overall configuration of the systems of innovation trajectories have varied.

In studying wave-like patterns in innovation trajectories, this chapter has introduced an explicit link between cyclical theories of technological evolution and the theories of innovation systems. In this context, the system perspective has helped to provide more insight into the dynamic processes of development (which were identified in Chapter 3). That is, although Chapter 3 illustrated how technologies in new technological systems builds upon old ones, this chapter illustrates how the scope and overall configuration in new systems still substitute (or creatively destroy) old ones.

The new conceptual and analytical challenges for the innovation system perspective, which this chapter has opened, will in Chapter 6 be integrated further with industry dynamics. In this context, the waves of innovation trajectories associated with system takeoff, identified in this chapter, will be related to the evolution of capabilities within industry structures, and *vice versa.*

6. Technological System Dynamics: A Competence Bloc Approach

Abstract

This chapter applies a capability perspective to the dynamics of industries. In this, the changing population of an industry system is derived from the changing distribution of capabilities of firms within industrial sectors, participating in the generation and exploitation of a common knowledge base. This competence bloc view on systemic change is an attempt to investigate the systemic aspects of innovation and competition at the microeconomic level. It is highlighted how industry systems evolve within technological structures, and *vice versa.*

The chapter illustrates how industrial sectors are coming together in adopting and developing technological trajectories within broader technological systems. This is in turn also reflected in the industrial sectors' broadening their technological competencies. This indicates an historical trend towards a more complex industrial base, and a more distributed future development of technologies. The results also open up a discussion of the limitations of the standard industrial classification scheme, if firms are to be grouped in relation to their closest 'dynamic' competitors, and why a system perspective is more appropriate.

6.1. INTRODUCTION: EXPLORING 'INDUSTRIAL DYNAMICS'

Studies within Industrial Dynamics have especially been in relation to competitive and co-operative processes, with respect to how firms create and respond (i.e. the behaviour of firms) to innovation (i.e. the role of innovation).[1] A major aim is to use this information to gain an understanding

[1] As mentioned in the Chapter 1 (section 1.3) one of the most comprehensive attempts to investigate the systemic aspects of innovation at the micro-economic level is provided by Carlsson and colleagues (1995, 1997). A central feature of their studies is

of the processes and performances of a specific socio-economic system (including its sustainability and its inherent contradictions). Among the several important contributions[2] made within Industrial Dynamics is the significance of the (unequal) distribution of economic competence and corporate capabilities (the ability to develop and exploit new business opportunities). The Industrial Dynamics literature also considers how systems can act as barriers as well as stimuli to knowledge accumulation. They make the important step of studying the systems supporting the development of technological trajectories as they evolve through several decades.

In this context this book takes the view that the changes in the configuration of innovation systems of technological trajectories (associated with wave-like patterns at takeoff or emergence), as identified in Chapter 5, do not happen in isolation, but in a dynamic interaction with its surrounding environment, including the industrial sectors and firms which act as the entities producing, modifying and using the component technological trajectories. In the Industrial Dynamics literature, it is believed that this interaction can have sweeping consequences for the structure of firms' and industrial sectors' technological competencies, the organisation of industry, as well as the further technological development.

The overall objective of this chapter is to examine and map the co-evolution and interaction between technological development and the industrial technological capabilities defining industry structures. The purpose of such information is to gain insight into the changing innovative and competitive landscape underpinning the dynamics of industry. Such insight is central to understanding the causes and effects of open-ended transformation processes of industry structures (i.e. the rules for changes).

The changing innovative and competitive landscape underpinning the dynamics of industry will be viewed from a capability perspective rather than the industry in relation to its product markets. This is in coincidence with the Industrial Dynamics literature, where focus has been on the underlying technological capabilities of firms relative to their major industrial competitors, or distributed capabilities of cooperative arrangements among firms (sometimes across the standard industrial classification schemes) and other institutions (at any level of aggregation), *rather* than the industry in relation to its product markets. Viewing industries from a capability perspective takes into account 'untraded interdependencies' (i.e. knowledge

---

the shift in emphasis from innovation to 'technological systems', and their conceptualisation around the notion of Industrial Dynamics.

[2] The Industrial Dynamics approach is much wider than the characteristics of it addressed within this chapter. For an overview on Industrial Dynamics, see Andersen and Lundvall (forthcoming).

interchange and interactive learning in parallel with, and often independently from, market transactions).

Within Industrial Dynamics the level of analysis is flexible, distributed across meso, macro and 'micro-based macro'[3] factors, which contribute to the development and transformation of a common knowledge (resource) base. The boundaries of Industrial Dynamics have especially been conceptualised in relation to system approaches (national innovation systems,[4] technological systems,[5] or distributed innovation systems[6]) or 'micro-based macro' models of the economy. Thus, the Industrial Dynamics literature typically involves a more interactive and multi-sectoral focus on change, in order to identify the industrial dynamic structures at a given time.[7] Hence, the units, levels and boundaries within Industrial Dynamics are determined by the particular problem of investigation, *rather* than the industry in relation to its product markets, and this is also the methodological starting point of this chapter.

For the purpose of the objective of this chapter (examining and mapping the co-evolution and interaction between technological development and the industrial technological capabilities defining industry structures) an interpretation of Carlsson's concept of technological systems is most applicable (Carlsson and Stankiewicz 1991, Carlsson 1997). In this approach systems of technological trajectories evolve within industry structures, and *vice versa*. The dynamic part of the system which is concerned with technology and industry selection, and which is an essential aspect of the path-dependent direction of the system, is based on an interpretation of Eliasson's concept of a 'competence bloc' (Eliasson 1996, 1997). In the

---

[3] See Chapter 1 (section 1.3) concerning the 'micro-based macro' (or 'micro to macro') perspective.

[4] Lundvall (1992) refers to 'national systems of innovation' in which he points to the existence of national specificities in 'collective entrepreneurship' as well as in user-producer interaction.

[5] Carlsson (1997) illustrates how innovative activity and industrial development co-evolve in relation to specific sectors and technologies

[6] Andersen, Metcalfe and Tether (2000) have suggested how emergence and shaping of innovative activity may be distributed across a wider matrix of structures, inter-relationships, constraints, and tradeoffs throughout the economic system.

[7] In this context the macro socio-economy is not simply the aggregate of various micro units, but is instead regarded as a complex outcome of micro relationships or interactions. Carlsson, Eliasson and Taymaz (1997) and Eliasson (1997) emphasise the micro to macro (or the 'micro-based macro') perspective as characteristic for Industrial Dynamics (See Chapter 1, section 1.3). This multi-sectoral focus is to be seen in contrast to the product cycle literature (Kuznets 1930, Vernon 1966, Utterback 1996, Utterback and Fernando 1993) which has helped us to understand changes in the characteristics of markets, technology and products along associated life cycles within specific industries.

'competence bloc' approach selection and competition is not constrained to product markets (as in mainstream reasoning). Long-term competitiveness is as much located within the firm's dynamic supply capabilities (e.g. technological capabilities, adjustment capabilities, local networks), or distributed in cooperative arrangements among firms and other institutions[8]. (For detailed description of the conceptual theoretical framework and terminology, see section 6.2).

The rationale of a system perspective in this chapter is to be able to foster an 'overview' of the historically changing structures of the nature of the innovative and competitive landscape, while not compromising on providing the details of the picture.

6.1.1. Chapter Outline

First, the data on technological development and industrial technological activity which form the basis for this chapter are selected. Then, the conceptual theoretical framework concerning the constituent parts of the technological systems are presented. The dynamic elements which capture the co-evolutionary process between technology and industry structures are also discussed. Special attention is given to 'competence bloc dynamics', uneven system development and the notion of emergence.

The chapter then turns to the empirical mapping of technological systems since the early interwar period. Building upon the historically changing innovation systems of technological trajectories (as identified in Chapter 5), this chapter empirically identifies the cross-roads (or co-evolution) of different industrial sectors' relative competencies with those systems, in order to empirically illustrate how the industrial dynamic competitive environment may have changed. However, for this purpose some conceptual and quantitative aspects with respect to how an 'industry system' can be measured. Some concluding remarks will follow in the final Section.

6.1.2. Data Selection

This chapter shares the view that takeoffs in innovation trajectories follow wave-like patterns of development. Thus the 98 Chemical, Electrical/electronics, Mechanical and Transport trajectories identified within such three waves (see Chapter 5, Tables 5.2-5.4) are appropriately used for

---

[8] The 'competence bloc' approach within Industrial Dynamics is compatible with the resource-based theories of the firm (Penrose 1959, Richardson 1972, Foss 1993). In the management literature similar issues have been addressed with respect to learning organisations.

the purpose of this analysis. The trajectories are based upon time series of accumulated patent stocks 1920-1990, and they are calculated at the 56 technological group level (Chapter 2). As in Chapter 5, this Chapter leaves aside Non-industrial technologies; this time due to the fact that there is no counterpart industrial sectors, by nature. The industries participating in Industrial Dynamics within this chapter are based upon the 20 industrial sector, and, as identified in Chapter 2, they are organised into 4 broad industrial groups: Chemical, Electrical/electronics, Mechanical and Transport. In the design of empirical categories and methods, further restrictions have been imposed when selecting corporate patenting for the analysis. They are presented, where appropriately, in the text.

6.2. THEORETICAL FRAMEWORK: TECHNOLOGICAL SYSTEM DYNAMICS

As mentioned in the Introduction (section 6.1), this chapter does not focus on organisational and institutional competencies, important though they are, but on the changing technological competencies of industrial sectors, responding to or interacting with the changing technological landscape and associated opportunities. The idea is that the an industry selects, uses and develops skills and competencies in an 'innovation system of technological trajectories' (as defined in Chapter 5); and that not only does the combination and scope of elements that go to make up any one such innovation system change over time, but so does the combination and degree of participation of the industrial sectors participating within it. An important issue is that not all technologies and industrial sectors participate equally in the evolving systems at any given time. However, a starting point here is that there is considerable overlap between the definition of an industry and of the technologies it uses.

The chapter builds upon the notion that, although a central characteristic of technological change is the way in which it gives rise to business and technological opportunities, sometimes in several industrial sectors, the ability and determination to exploit those opportunities is associated with economic and corporate competence, which is firm-specific, cumulative in character and often tacit. It is proposed that industrial sectors differ, just as firms do (see Chapter 7), with respect to the influence on their dynamic technological capabilities from past collective learning processes, inherited routines and expertise.[9] For industrial sectors this is also related to their

[9] Based upon the pioneering works of Rosenberg (1976, 1982) and Nelson and Winter (1982) emphasis within Industrial Dynamics is on how technologies, organisations and the dynamic capabilities of firms change only incrementally and are

product specialisation and historical role. That is, on the one hand the component technological trajectories provide opportunities for different industrial sectors, but on the other hand those opportunities will not be equally appropriate to the different industrial sectors or consistent with the strategies to be found in their constituent firms.

Accordingly, the evolution of industrial technological competence is a function of the changing technological landscape and associated opportunities, what the collection of firms in an industrial sector need (or decide they need) to implement the strategies they have adopted, and their capabilities of doing so. In this framework, firms only optimise locally, if at all, (i.e. there is no generally applicable best corporate practice), since they are constrained in the choices they make and by the resources and capabilities that they control (everything is context specific). Hence, the mainstream economics production function cannot be taken for granted.

6.2.1. The Technological System

This analysis (of examining and mapping the co-evolution and interaction between (i) technological development and (ii) the industrial technological capabilities defining the dynamic structures of industry) needs a theoretical organisation that captures the technology-industry co-evolutionary process in sufficient detail, and, as described in the introduction, an appropriate starting point is that of technological system dynamics defined along an interpretation of Carlsson (Carlsson and Stankiewicz 1991, Carlsson 1997).

Technological systems are here distinguished from the notion of national innovation systems, which is mainly based upon national innovation institutionalism (Freeman 1987, Nelson 1992) or national specificities in 'collective entrepreneurship' (Lundvall *et al* 1992). Whereas a national innovation system includes a country's firms and supporting institutions (education system, research organisations, financial institutions, culture, government policies, and so on) which may enable, for example, limited resources to be efficiently used in innovation or may cause abundant resources to be squandered (Freeman 1987); a technological system, along the lines presented by Carlsson and Stankiewicz (1991) and Carlsson (1997,

---

cumulated along specific paths or trajectories. Nelson (1991a) reviewed how the boundaries to firms' behaviour also explain why firms differ in their knowledge base, in their capability to innovate, and in their ability to appropriate from the innovations they make. Hence, the mainstream notion of the representative firm does not apply. (See further Chapter 7 which develops a taxonomy of types of corporate technological competencies, and subsequently investigates how the positions of firms matter for their future trajectories.)

1998), moves beyond the nation state and focuses on knowledge concerning technical matters in relation to industrial networks and economic development. However, whereas Carlsson and Stankiewicz's (1991, p.93) applied their notion of technological system dynamics mainly in relation to industrial districts or other geographical phenomena, this chapter applies it in terms of an historical epoch or 'wave' governing technological and business opportunities, a use more closely related to the way Freeman and Perez (1988) apply the term 'technology system', though this is not a central feature of their analysis, which focuses more on techno-economic paradigms and technological revolutions.

For analytical purposes this chapter sub-divides a technological system into three co-evolving sub-systems: (i) An 'innovation system of technological trajectories (as identified in Chapter 5)', (ii) An 'industry system' and (iii) 'Other systems' (e.g. technological support systems including universities, science parks etc.). However, this chapter is only concerned with the co-evolution of (i), the 'innovation system of technological trajectories', and (ii) the 'industry system'. The participation and effect of (i) and (ii) are embedded in (iii), other systems.

In this context, the idea behind the co-evolutionary perspective is that, the identification of the industrial sectors' changing technological competencies cannot be separated from their co-evolution with the technological trajectories themselves. That is because technological development in different industrial sectors creates variety in the competencies of those sectors. At the same time, changes in technological expertise and routines only occur gradually and within certain boundaries over time. As a consequence, each industrial sector becomes locked into particular paths of development. It is believed here that the specific collection of technological competencies of any industrial sector is likely to influence both the form and the future direction in which the spectrum of that industry's technological competencies evolve. At the same time it will influence the future direction of the development of the component technological trajectories.

Thus, an 'innovation system of technological trajectories' is inseparable from the skills, competencies, strategies, learning abilities, day-to-day practices, rules of thumb, and definitions of - and approach to - interesting problems of a given collection of firms and their industrial sectors (i.e. the 'industry system') at any particular moment.[10]

---

[10] This is because the nature of the skills, competencies, routines, etc. (associated with the 'innovation system of technological trajectories') that a particular collection of firms and industries explore, develop and exploit change over time. At the same time the collection of firms and industries (associated with the 'industry system') using the shifting or changing 'innovation system of technological trajectories' will also change

Accordingly, the 'innovation system of technological trajectories' co-evolves with the 'industry system' in an open-ended process of continuous change and adaptation. But for analytical purposes only, this chapter considers them separately.

This is, of course, a very stylised picture of a technological system and how it operates, but it is done for analytical purposes in order to measure empirically the co-evolution between them.

(i) The 'innovation system of technological trajectories': Defining the term.

The innovation system concept here ought to be understood in a very *neutral* way, as an 'innovation system of technological trajectories'. Hence, the innovation system concept here is very different from the one used in the national systems of innovation literature, as described above.

The definition of an 'innovation system of technological trajectories' is here a collection of related product, process and application technologies, whose composition, boundaries, opportunities and relative importance change over time (as identified in Chapter 5), as does the dynamic interaction of the system with industrial sectors and their firms. The innovation system of technological trajectories is made up by many component technological trajectories.

An 'innovation system of technological trajectories' does not have clearly defined boundaries, but, in a strictly technological sense, when there is a collection of technological innovation trajectories with an interrelated collective scope interacting at the 'technological frontier', then they serve as the base for the technological boundaries (or system component technological trajectories) in the measurement of the system. Not only are the technological boundaries (or system component technological trajectories) a part of defining the configuration of the innovation system, but so is the scope of each of the individual technological trajectories (or component technological trajectories) co-evolving within it.

The scope of an 'innovation system of technological trajectories' is related to:

- the structure of opportunities of the technological trajectories (or component technological trajectories) evolving within it (measured by

over time, due to their dynamic technological capabilities from past collective learning processes, inherited routines and expertise which change only incrementally over time. (See section 6.2.2 on competence bloc dynamic) the percentage rate of growth in accumulated patent stock over the trajectory growth cycles).

- the structure of technological core importance, c.f. socio-economic competence or historical importance, of the technological trajectories evolving within it. (Measured by the accumulated stock of patent stock, at the point of takeoff of the trajectory growth cycles).

The main carriers of the 'innovation system of technological trajectories' are then defined as those with the highest opportunities and of highest core importance.

The technological trajectories forming the base for this chapter, are derived in Chapter 4, and the associated innovation systems of technological trajectories (including their scope and overall configuration) are derived in Chapter 5. The component technological trajectories of those innovation systems are further listed in Tables 6.1-6.3 (together with the industrial sectors co-evolving with them).

To set an illustrated example, Figure 6.1. (see section 6.2.2.) pictures such an innovation system of three technological trajectories combined within an ellipse.

(ii) The 'industry system': Defining the term.

In line with Industrial Dynamics, a central feature of the technological system is the dynamic capabilities of firms (the ability to develop and exploit new business opportunities). In the economic literature firms and industries are also increasingly grouped in relation to their technological competencies (or dynamic capabilities) which in an evolutionary context have an important influence on their long-term survival and performance. For example, Walsh (1997) and Andersen and Walsh (2000) define the broad chemical industry as 'the complex of industries related to the exploitation of chemical knowledge or which have grown around a chemical base'.

Accordingly, an 'industry system' is here defined as the collection (or block of firms) within industrial sectors with relatively large technological capabilities to innovation in at least one of the component technological trajectories of an 'innovation system of technological trajectories' (as defined above). Hence, an 'industry system' is defined in terms of a corporate capability base surrounding an 'innovation system of technological trajectories', rather than in relation to product markets. Figure 6.1 (see section 6.2.2.) illustrates such an 'industry system' of three industrial sectors, combined within an ellipse, who exploit the components of an 'innovation system of technological trajectories' as input into their problem solving.

Building upon this approach, and defining an 'industry system' as the complex of industrial sectors having accumulated a competence in developing and exploiting the system of component technological trajectories at a given time, also imply that we do not define either the 'industry system' or the 'innovation system of technological trajectories' independently of one another. It also implies, explicitly, that the 'industry system' is defined in relation to corporate capabilities rather than product markets, which is very central to the notion of Industrial Dynamics (see section 6.1). Finally it implies that the 'industry system' does not have any fixed boundaries as industries grouped within standard industrial classification schemes.

6.2.2. Competence Bloc Dynamics

Drawing upon Eliasson (1997, p.5), the part of the system which is concerned with selection and choices (picking technologies and corporate winners, removing technologies and corporate losers) we call the 'competence bloc', as also elaborated upon earlier (see section 6.1). This is set into motion by the experimentally organised economy (Eliasson 1991). However, in this chapter the 'competence bloc' is not related to industrial districts, as in Eliasson (1996, 1998), but it is rather an industrial competence bloc located in a group of firms and industrial sectors sharing a common technology resource base.

A visual interpretation of this approach is given in Figure 6.1, in which a 'competence bloc' is the combination (or crossroads) of the 'innovation system of technological trajectories' and the matching 'industry system' at a given time. Figure 6.1 is a particular attempt to use the competence bloc approach to capture the dynamics of the technological system, especially concerning the selection and evolution of choices and technological opportunities, and the firms and industries developing and exploiting them. Using competence bloc dynamics as an explicitly analytical tool (in the way applied here) was developed by Andersen and Walsh (2000).

A central point here is that as technology is assumed to be differentiated between firms, industrial sectors and nations, the industrial sectors with the most appropriate technological capability (i.e. whose organisational structure and routines provide the best fit with the component technological trajectories of the innovation system governing at this time) survive and thrive in a process of natural selection (in the way described by Nelson and Winter 1988, Perez 1989). However, this model also allows for learning and adjusting as a way of surviving, so developing and building competencies in the 'right' technological sectors becomes an essential part of the dynamic competition among firms and industries.

Figure 6.1 Technological system dynamics: a 'competence bloc' approach

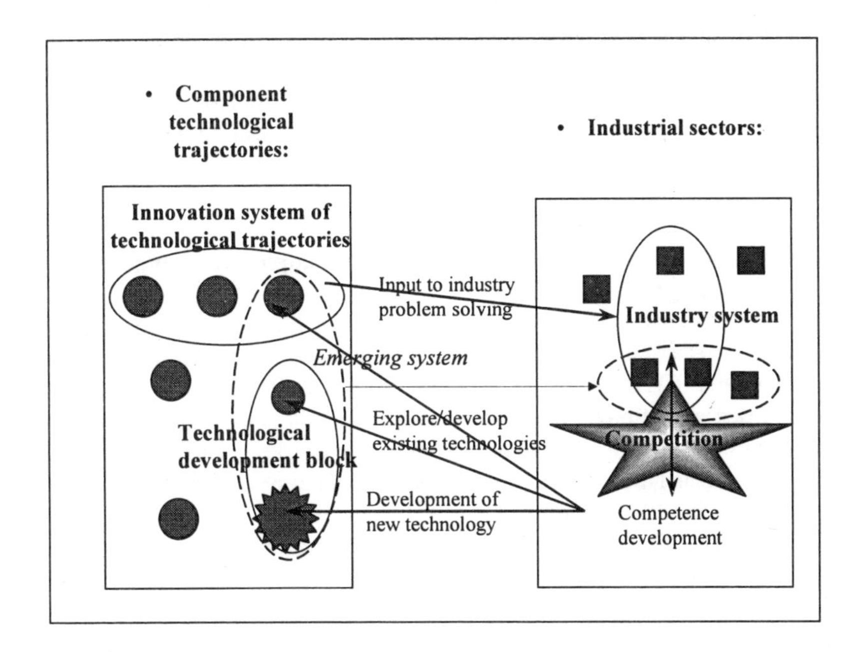

Source: Revised version of Andersen and Walsh (2000).

Similarly, just as a pervasive new 'innovation system of technological trajectories' (or system of component technological trajectories in Figure 6.1) selects the best-adaptable environment (firms and industries) for diffusion, the firms and industries in that environment also contribute to, and thereby select, the rate and direction of technological activity. As continuous technological exploration and development comes from, and adds to, the complex of component technological trajectories, as well as complementary technological developments existing elsewhere (see 'Technological development block' in Figure 6.1), and innovations build on each other, technological development can be regarded as being cumulative. Technologies with entirely new features of course also appear from time to time (see star-like component technology in 'Technological development block' in Figure 6.1). In similar ways, firms and industries accumulate their competencies as well as develop new ones.

When firms and industries develop competencies in new technologies, outside the existing system of component technological trajectories, this can have an overall competence-destroying effect for the existing competence bloc, as it creates mismatch. A competence-destroying effect within this framework is regarded as creative in the sense that it boosts a new system (see emerging system of component technological trajectories and industrial sectors captured by emerging ellipses). Mismatches are the dynamic force for potential new developments and change. The notion of emergence mismatches (or uneven system development) are further elaborated upon in Section 6.2.3. below. Some technologies, firms or industrial sectors may of course also survive or die in the selection process because agents (e.g. from the support system) are able to change the selection environment.

6.2.3. Uneven System Development and Development, Change and Emergence

Since there is assumed to be no single technological system movement at the frontier, but (i) different movements of different generic technological trajectories at different rates and socio-economic impact, as well as (ii) different degrees of involvement of different industries at different times, the above described configuration of a technological system imply an uneven type of system development.[11] This also suggests that dynamic force within the formation of broader technological systems is created by internal tensions in which economies of scope can be achieved from the accumulation, catching up and combination of different component technological trajectories, co-evolving with changing industry structures and boundaries.

As mentioned in Chapter 1 (section 1.3) such ideas were also presented by Hughes (1992), who refers to salients (or reverse salients) in the development phase and to the solving of critical problems. Hughes also points out that some components or sets of components in a technological system may lag behind (or advance in front) of a system as a whole, because they are less (or more) efficient or economical, or in some other way show up in the backwardness (or advanced state) of the other components in the system. An impetus is then given to the system as a whole, in which the most efficient technologies create the development potential for the others. The salient metaphor and the notion of uneven technological system development can also be related to Dahmen's (1989) concept of development blocks (a term which is adopted in this chapter, see Figure 6.1) and Dosi's concept of technological paradigms and the evolution of technological trajectories (Dosi 1982, Dosi *et al.* 1988). Rosenberg (1982) writes of bottlenecks in technological development to capture a similar idea to Hughes's 'reverse-

[11] Unsynchronized technological development was also confirmed in Chapter 4 and 5, with respect to scope, impact, phases of development, and takeoff points (although grouped within certain time periods) of different technological trajectories.

salients'. Rosenberg also uses the notion of 'focusing devices' in arguing that technological change historically moves in 'patches'

Furthermore, as technologies and industrial sectors' corporate technological capabilities grow at different rates and with different impacts at different times, we particularly expect a change in structure at the technological frontier historically. This process, in its turn, leads to new combinations and sometimes the emergence of new technological systems. Co-evolution, therefore, is not only about catching up and accumulation, but also about new combinations of technologies and industry structures, and the breaking down of old ones (which relates to Schumpeter's ideas about creative destruction). This is also the idea, for example, behind Kodama's notion of technological fusion and the outgrowth of new industries (Kodama 1986, 1992), and behind the work of Carlsson and Stankiewicz (1991) on the nature, function and composition of technological systems.

Hence, the boundaries, scope and overall configuration of the technological system is important for the system's pattern of development, and the systems can act as barriers as well as stimuli to knowledge accumulation.

6.3. CO-EVOLVING WITH 'INDUSTRY SYSTEMS'

The analysis now turns to empirical measurement issues with respect to identifying the 'innovation system of technological trajectories' as well as the 'industry system', in order to subsequently examine and map their co-evolution. The changing 'innovation systems of technological trajectories' (associated with waves of system development) were measured and identified in Chapter 5. The configuration (boundaries and scope) of the 'innovation systems of technological trajectories' are presented in Tables 5.2 to 5.4

Hence, we now consider the interaction of this evolution with changing industry structures. In measuring the changing 'industry systems' the study will, in accordance with the technological dynamics perspective from section 6.2.1, especially focus on the boundaries and structures of the industries' technological profiles (or competencies) in the component technological trajectories of the changing 'innovation systems of technological trajectories'.

6.3.1. The 'Industry System' and How it Can Be Measured

It is now commonly accepted that the pattern of technological specialisation serves as an indicator for technological competence or strength at various levels of aggregation, so that corporate patents at the level of industrial sectors can be used as a proxy for the underlying pattern of industrial technological competencies. Patent data are particularly appropriate for this type of measure, as the classification scheme captures the structural elements of their technological specialisation, in a way that e.g. research and development (R&D) expenditures for example do not, and corporate patents are recorded historically, whereas formalised R&D laboratories (with associated R&D budgets) are a more recent phenomena. (See also Chapter 2 on appropriate indicators.)

Using the accumulated patent stock for each industrial group, each industry's pattern of technological specialisation is measured by an index termed the Revealed Technological Advantage (RTA). The RTA index of an industry in a particular technology is here given by the industry's share in a technological group of US patents, relative to the overall share of all US patents granted to that industry out of the total. Hence, the RTA index for each industry in each technology group is defined as: $RTA_{ij} = (P_{ij}/\Sigma_i P_{ij}) / (\Sigma_j P_{ij})/(\Sigma_{ij} P_{ij})$, where P_{ij} is the number of US patents granted to industry i in technological group j. The index varies around unity, such that values greater than one suggest that an industry is positively or comparatively specialised in technological development in the technological group in question.

The RTA_{ij} index has been calculated for all industrial sectors (as identified in Section 6.1.2) in all Chemical, Electrical/electronic, Mechanical and Transport 56 technological groups (as identified in Chapter 2) for every five years from 1920 to 1990.

The pattern of industrial sectors' technological competencies in the shifting systems of component technological trajectories, whatever configuration they may have, can now be measured in order to establish empirically the changing boundaries and overall configuration of the Chemical, Electrical/electronic, Mechanical and Transport 'industry systems', including their boundaries.

When identifying the structure of industrial technological competencies during the evolution of 'innovation systems of technological trajectories', the degree to which an industry specialises in a component technology at the point of system takeoff is important, as this can provide us with information concerning the broader features of the system. It is here believed that the structure of the technological competencies of the industrial sectors launching the system is likely to influence (i) the form and the future direction of the system of component technological trajectories' development, as well as (ii) the direction and structure of the spectrum of the industry's technological competencies. That is, as changes in technological expertise and routines only occur gradually and within certain boundaries over time, so that industrial sectors have become locked into certain specific paths of development, the distribution of competencies within the component technological trajectories of systems and industrial sectors at the point of takeoff tells something about the industrial sectors which launched the system. In Chapter 8 it is also confirmed that a high rate of technological development is associated with a high rate of diffusion, so by focusing on the industrial sectors participating at an early stage of the growth curve we capture the industrial sectors' launching the system and thereby its characteristics. Hence, although it is expected that a technological system provides opportunities for a broader spectrum of industrial sectors when it evolves as well as grows (as its component technological trajectories diffuse and play an important socio-economic society part); the features, opportunities and problem solving activity within the system still reflect the complex of related industrial sectors associated most closely with the emergence of the system.

Measuring the configuration of the 'industry system' (including its boundaries)

As mentioned in section 6.2.1, the 'industry system' (or industrial boundaries) is here defined as the collection (or block) of firms within industrial sub-sectors with relatively large competencies in at least one of the technological components in the 'innovation system of technological trajectories'. This is measured by the value of the RTA index for each of the industrial sub-sectors in each of the component technological trajectories within the chemical, electrical/electronic, mechanical or transport innovation systems at the time of its unleash (or takeoff). If $RTA_{ij}>1$ for an industrial sub-sector in a component technology of the innovation system at the point of takeoff, this means that the industrial sector is participating actively in the system.

Hence an 'industry system' can be very simple (i.e. made up of few industrial sectors having competencies in few component technological trajectories), or it can be very complex (i.e. a broad range of industrial sub-sectors having competencies in a broad range of component technological trajectories), or somewhere between the two. In this approach, several industrial sectors can also participate in several innovation systems of technological trajectories, and thereby play a role in the dynamics of several broad 'industry systems' (i.e. Chemical, Electrical/electronic, Mechanical and Transport broad industries). In that way, it is illustrated empirically how the industry system boundaries are flexible, as well as change over time.

6.4. EMPIRICAL MAPPING

Based upon the RTA_{ij} of all industrial sectors calculated across all Chemical, Electrical/electronic, Mechanical and Transport technological trajectories belonging of innovation systems, the crossroads between the 'innovation systems of technological trajectories' and the industrial sectors' competencies in those trajectories around the year of takeoff (i.e. the 'industry system') are presented in Tables 6.1 to 6.3.

However, with respect to the industrial sectors presented in Chapter 2, and for the purpose of this analysis, the industries of Tobacco (Industry 3) and Other Manufacturing (Industry 20) are left out. Furthermore, the Food industry (Industry 1) has been combined with the Drink industry (Industry 2); Aircraft (Industry 11) with Other Transport Equipment (Industry 12); and Paper Products (Industry 14) with the Printing and Publishing industry (Industry 15). This was to avoid problems with few firms resulting in small numbers of patents, which would cause only random results. Hence, each organised industrial sector has at least 50 patents (and for the most – significantly more) at the calculation of their RTA indexes at various times.

The RTA_{ij} index calculated for the fifth year period closest to the takeoff point has been used as the point of reference, except in cases where the duration of the technological trajectory is less than 10 years. In those cases the fifth year period before its takeoff is always selected, in order to capture the industrial sub-sectors' competence in the technological trajectory before it has moved into prosperity and possibly diffused into a broader spectrum of industrial sectors.

The combination and underlying technological and industrial configurations of the crossed boxes within Tables 6.1 to 6.3 (i.e. in which $RTA_{ij} > 1$), illustrate the complex boundaries of what in this chapter has been termed or denoted a 'competence bloc'. In Figure 6.1 (section 6.2.2) this is also illustrated by the combination of the components in the 'innovation system of technological trajectories' and the 'industry system' at a given time.

Four such competence blocs are presented across the three waves (identified in Chapter 5) of system unleash. The four competence blocs include the Chemical, Electrical/electronic, Mechanical, and Transport innovation systems of technological trajectories. See Tables 6.1 to 6.3.

The results indicate, perhaps not surprisingly, that the takeoff of an 'innovation system of technological trajectories' generally tends to be launched by the industrial sector most related to the technology indicated by the specialisation (or high RTA_{ij}) in that sector. Thus the rubber and plastics products industrial sector is highly specialised in the technology of making rubber and plastics products; professional and scientific instrument firms (such as camera manufacturers, and including Eastman Kodak which has since de-merged into Kodak and Eastman fine chemicals) were specialised in photographic chemicals; food firms specialise in food; the non-metallic mineral products industrial sector (mainly glass and ceramic firms) are heavily specialised in the technology of making non-metallic mineral products; coal and petroleum products firms are highly specialised in the technology of making coal and petroleum products; pharmaceuticals firms are highly specialised in pharmaceuticals and biotechnology; metal firms are highly specialised in technologies related to metallurgical processes and miscellaneous metal products etc. However, it was less so over time.

Whereas evidence for the interwar technological system (or 'Electrical and heavy engineering [third] Kondratieff' (Chapter 5)) showed a relative large overlap between industry and technology (i.e. chemical industries tend to specialise in chemical technologies and so forth) except for transport which are also heavily specialised in mechanical technologies related to metals and instrument and controls; the postwar technological system (or 'Fordist mass production [fourth] Kondratieff' (Chapter 5)) was a much more complex system with extended technological and industrial boundaries, both in terms of the total number of component technological trajectories making up the system, and in terms of the interrelated overall scope of those component technological trajectories. This also reflects the evidence in Chapter 3, which characterises the interwar period by intra-group technological diversification and the formation of a structure of specialised engineering and science-based fields (as opposed to more complex structures of trajectories), whereas the movement from the postwar period towards recent time is characterised by inter-group technological concentration and the formation of a structure of interrelated engineering and science-based fields.

Finally, the book forecasts or speculates an emerging system (or fifth 'Information and communication Kondratieff', (Chapter 5)) which moves us into the life sciences and the exploitation of biotechnology. So far this technological system seems very 'concentrated', with respect to Electrical and Mechanical industries specialising in the component technological trajectories of their respective innovation systems. However, Chemical industries diversify their specialisation across several broad innovation systems of technological trajectories, and Transport industries tend to specialise in the Mechanical 'innovation system of technological trajectories'. Whether this technological system (which is still just at the emerging phase) is going to be a combination of a semi-concentrated (or narrowly specialised) and a semi-complex (diversified and technological interrelated, combining several different knowledge bases in its evolution), or whether it will follow the historical trend towards more complex technological systems in general, is still early days to tell.

Table 6.1. Technological system dynamics: The competence bloc of "Wave I"

Revealed Technological Advantage for industry i in technological group j (RTAij) around year of technological takeoff: xxx: RTA > 3; xx: 2 < RTA < 3; and x: 1< RTA < 2		The closest fifth year around takeoff	Chemicals n.e.s.	Pharmaceuticals	Textiles and clothing	Coal and petroleum products	Electrical equipment n.e.s.	Office equipment	Food / Drink	Metals	Mechanical engineering n.e.s.	Paper products n.e.s./ Printing and publishing	Non-metallic mineral products	Professional and scientific instruments	Motor vehicles	Aircraft / Other transport equipment	Rubber and plastic products	Coefficient of variation (CV) across all RTAij in a technological group
			Chemical				Electrical/ electronic		Mechanical						Transport			
Code	Technological group j:	INDUSTRY i:	4	5	13	18	8	9	1	6	7	14	17	19	10	11	16	CV
CHEMICAL																		
2	**Distillation processes**	**1930**	**xxx**			**xxx**								**xx**				**226.52**
3a	**Inorganic chemicals**	**1930**	**xxx**		**x**													**198.19**
4a	Agricultural chemicals	1930	xxx						xxx									259.71
5a	**Chemical processes**	**1930**	**xx**		**x**	**x**							**x**				**xxx**	**184.05**
7a	**Cleaning agents and other compositions**	**1930**	**xxx**		**xxx**	**xxx**							**x**	**xxx**				**158.28**
8a	Disinfecting and preserving	1925	xxx	xxx		xxx				x								257.34
10a	**Bleaching and dyeing**	**1930**	**xxx**		**xxx**													**315.83**
51a	**Coal and petroleum products**	**1930**	**x**			**xxx**												**339.17**

Code	Technological group j:	INDUSTRY i:	4	5	13	18	8	9	1	6	7	14	17	19	10	11	16	CV
55a	Explosive compositions and charges	1930	xxx															374.96
ELECTRICAL / ELECTRONICS																		
30a	Mechanical calculators and typewriters	1920						xxx										381.89
33	**Telecommunications**	**1925**					**xxx**											**372.13**
34	Other electrical commun ication systems	1935					xx	x										178.79
36a	Image and sound equipment	1925					xx								x			241.18
37a	**Illumination devices**	**1925**											**x**		**x**			**147.75**
38a	**Electrical devices and systems**	**1920**					**xx**											**223.89**
39a	**Other general electrical equipment**	**1925**					**x**											**138.22**
52a	Photographic equipment	1920												xxx				386.86
MECHANICAL																		
la	Food and tobacco products	1930	x	xxx					xxx	xx								331.21
13a	**Metallurgical processes**	**1925**	**x**				**x**			**xxx**					**xx**	**xxx**		**110.48**
14a	**Miscellaneous metal products**	**1920**							**xxx**	**xxx**			**xxx**	**x**	**xx**	**xx**		**131.20**
16a	Chemical and allied equipment	1925	x	x	x	xxx			xx	x	x	xxx	xxx			xxx	x	90.04
18a	Paper making apparatus	1930	x	xx		xxx			xxx	xxx		xxx						247.59
19a	Building material processing equipment	1925									x	xxx	xxx					240.60
22a	Other construction and excavating equipment	1925									xxx							358.88

Code	Technological group j:	INDUSTRY i:	4	5	13	18	8	9	1	6	7	14	17	19	10	11	16	CV
22b	Other construction and excavating equipment	1930								xxx	xx			xxx				220.42
23a	**Mining equipment**	**1925**	x			xxx				xxx								**327.47**
24a	**Electrical lamp manufacturing**	**1930**	xx				xxx											**260.78**
25	Textile and clothing machinery	1925									xxx							237.19
28a	Other specialised machinery	1925			xx	x			x	x	xx	x					xxx	110.50
29a	**Other general industrial equipment**	**1925**				x					x		xx		x			**92.87**
50a	Non-metallic mineral products	1925	x	x		x			x	x		xxx	xxx		x	x	x	155.76
53a	**Other instruments and controls**	**1925**					x	xxx		x		x		xxx	x	xx		**113.89**
TRANSPORT																		
44	**Aircraft**	**1925**						x							xxx	xxx	xxx	**266.24**
45a	Ships and marine propulsion	1930						xxx		xxx					xx	xxx		144.10
49a	**Rubber and plastic products**	**1920**	x		xxx	x									x		xxx	**272.75**

Note: The technological main carriers of the system are highlighted in bold

Table 6.2. Technological system dynamics: The competence bloc of "Wave II"

Revealed Technological Advantage for industry i in technological group j (RTAij) around year of technological takeoff: xxx: RTA > 3; xx: 2 < RTA < 3; and x: 1< RTA < 2		The closest fifth year around takeoff	Chemicals n.e.s.	Pharmaceuticals	Textiles and clothing	Coal and petroleum products	Electrical equipment n.e.s.	Office equipment	Food / Drink	Metals	Mechanical engineering n.e.s.	Paper products n.e.s./ Printing and publishing	Non-metallic mineral products	Professional and scientific instruments	Motor vehicles	Aircraft / Other transport equipment	Rubber and plastic products	Coefficient of variation (CV) across all RTAij in a technological group
			Chemical				Electrical/ electronic		Mechanical						Transport			
Code	Technological group j:	INDUSTRY i:	4	5	13	18	8	9	1	6	7	14	17	19	10	11	16	CV
CHEMICAL																		
3b	Inorganic chemicals	1965	xx			x				x		x	x					101.17
4b	**Agricultural chemicals**	**1970**	**xx**			**x**			**x**								**x**	**128.21**
5b	**Chemical processes**	**1965**	**x**		**x**	**x**				**x**		**x**	**x**				**xx**	**60.20**
6	**Photographic chemistry**	**1965**	**xx**					**x**						**xxx**				**307.64**
7b	Cleaning agents and other compositions	1960	x		x	xxx			x					x				136.11
7c	**Cleaning agents and other compositions**	**1970**	**x**	**x**		**xxx**			**x**									**117.29**
8b	Disinfecting and preserving	1970	x			xxx			xxx		x	xxx	x					160.91

Code	Technological group j:	INDUSTRY i:	4	5	13	18	8	9	1	6	7	14	17	19	10	11	16	CV
9	**Synthetic resins and fibres**	**1960**	**xx**		**x**	**x**			**x**					**x**			**xxx**	**137.69**
10b	Bleaching and dyeing	1970	xxx	x	xxx													212.65
51b	Coal and petroleum products	1955				xxx												321.68
51c	Coal and petroleum products	1965				xxx												313.86
55b	Explosive compositions and charges	1965	xx			xx										x		183.77
ELECTRICAL / ELECTRONICS																		
30b	Mechanical calculators and typewriters	1970						xxx				x						286.65
35a	**Special radio systems**	**1945**					**xx**											**299.41**
36b	Image and sound equipment	1975					xxx	x								x		152.68
37b	Illumination devices	1955					xx											223.07
38b	**Electrical devices and systems**	**1945**					**xx**											**195.37**
38c	**Electrical devices and systems**	**1965**					**xx**	**x**										**136.08**
39b	Other general electrical equipment	1945					x	xx					x		x			133.28
39c	**Other general electrical equipment**	**1965**					**x**	**x**		**x**			**x**		**x**	**x**		**89.87**
40	**Semiconductors**	**1965**					**xx**	**xxx**										**222.22**
41	**Office equipment and data processing systems**	**1975**					**x**	**xxx**										**206.90**
52b	Photographic equipment	1945						x				xx		xxx				313.97
52c	Photographic equipment	1970	x					xx						xxx				295.40

Code	Technological group j:	INDUSTRY i:	4	5	13	18	8	9	1	6	7	14	17	19	10	11	16	CV
MECHANICAL																		
1b	Food and tobacco products	1965	x	xxx					xxx	x		xxx		x				270.68
13b	Metallurgical processes	1965					x			xxx		x			x	x		112.12
15a	Food. Drink and tobacco equipment	1950					x		xxx	x	xxx		x					225.34
15b	Food. Drink and tobacco equipment	1955					x		xxx	x	xxx		x					212.75
15e	Food. Drink and tobacco equipment	1965					x		xxx		xxx							225.67
16b	Chemical and allied equipment	1950			x	xx			x	x	x	xx	xxx		x			93.19
16e	Chemical and allied equipment	1965			x	x			x	x	x	xx	xxx		x			73.48
17a	Metal working equipment	1970			x					xxx	xx	xx	x		x	x	x	89.65
18b	Paper making apparatus	1965	x	x	xx				xxx	xxx	x	xxx						193.43
19b	Building material processing equipment	1965			x					x	xx	xx	xxx		x			242.41
20	Assembly and material handling equipment	1960			xxx			x	x	xxx	xx	xxx	x	x				87.73
21a	Agricultural equipment	1950									xxx							369.58
21b	Agricultural equipment	1965									xxx							330.62
22c	Other construction and excavating equipment	1965									xxx				x			309.85
23b	Mining equipment	1960								xxx								267.43
24b	Electrical lamp manufacturing	1965					xxx											333.10
26a	Printing and publishing machinery	1965			xxx		x				xxx	x					x	186.31
26b	Printing and publishing machinery	1970			x			xxx		x	x	xxx	x	x				121.91
27a	Woodworking tools and machinery	1955							xx		xxx	xxx						318.26

Code	Technological group j:	INDUSTRY i:	4	5	13	18	8	9	1	6	7	14	17	19	10	11	16	CV
27b	Woodworking tools and machinery	1970			xxx					x	xxx	xxx	x					289.61
28b	Other specialised machinery	1950			x				x	xxx	xx	xx	x		x	x	xxx	73.21
28c	**Other specialised machinery**	**1965**			**x**				**x**	**xx**	**xx**	**x**	**xx**		**x**	**x**	**xx**	**53.63**
29b	**Other general industrial equipment**	**1965**						**x**			**xx**				**xxx**	**xxx**		**109.39**
31	**Power plants**	**1955**						**x**			**x**				**xxx**	**xxx**		**185.01**
50b	**Non-metallic mineral products**	**1970**	**x**		**xx**					**x**		**xx**	**xxx**				**x**	**107.62**
53b	**Other instruments and controls**	**1965**					**x**	**x**				**x**		**xx**	**x**	**x**		**63.45**
TRANSPORT																		
42	**Internal combustion engines**	**1980**								**x**					**xxx**			**276.42**
43	Motor vehicles	1965								xx					xxx			245.14
45b	Ships and marine propulsion	1965				x		x	x	xx					x	xxx	xxx	114.63
46	Railways and railway equipment	1970								xxx	x	xx	xxx					205.43
49b	Rubber and plastic products	1965	x		xxx								xxx	x			xxx	159.73

Note: The technological main carriers of the system are highlighted in bold

Table 6.3. Technological system dynamics: The emerging competence bloc of emerging Wave III (necessarily speculative)

Revealed Technological Advantage for industry i in technological group j (RTAij) around year of technological takeoff: xxx: RTA > 3; xx: 2 < RTA < 3; and x: 1< RTA < 2		The closest fifth year around takeoff	Chemicals n.e.s.	Pharmaceuticals	Textiles and clothing	Coal and petroleum products	Electrical equipment n.e.s.	Office equipment	Food / Drink	Metals	Mechanical engineering n.e.s.	Paper products n.e.s./ Printing and publishing	Non-metallic mineral products	Professional and scientific instruments	Motor vehicles	Aircraft / Other transport equipment	Rubber and plastic products	Coefficient of variation (CV) across all RTAij in a technological group
			Chemical				Electrical/ electronic		Mechanical						Transport			
Code	Technological group j:	INDUSTRY i:	4	5	13	18	8	9	1	6	7	14	17	19	10	11	16	CV
CHEMICAL																		
12	Pharmaceuticals and biotechnology	1990	xx	xxx														231.68
ELECTRICAL / ELECTRONICS																		
30c	Mechanical calculators and typewriters	1985					x	xxx										279.38
35b	Special radio systems	1985					xxx	xxx								xx		163.86
38d	Electrical devices and systems	1985					xx	x							x			121.66

Code	Technological group j:	INDUSTRY i:	4	5	13	18	8	9	1	6	7	14	17	19	10	11	16	CV
MECHANICAL																		
1c	Food and tobacco products	1985	x	xx					xxx			x						310.25
14b	Miscellaneous metal products	1985			xx				xx	xx	x	xxx	xx		xx	x	x	65.09
17b	Metal working equipment	1985							x	xxx	xxx	x	x		x	x	x	101.15
29c	Other general industrial equipment	1985					xx				x		x		xxx	xx		105.66
TRANSPORT																		

Hence, defining an industry around the concept of an 'industry system' in terms of a corporate capability base surrounding an 'innovation system of technological trajectories', we can see how industries move into various broad innovation systems of technological trajectories. This makes the broad industry system's boundaries more blurred. Hence, just as the nature of the innovation systems of technological trajectories have changed in boundaries and scope over time, in which they have become more interrelated and complex, so has the complex of industrial sectors making up the broad Chemical, Electrical/electronic, Mechanical and Transport 'industry systems' broadly defined.

The overall effects the dynamics of the technological systems may have had on the boundaries of the 'industry systems' can also be measured by the changing distribution, or spread, of industrial sectors' relative specialisation in the component technological trajectories of innovation systems. To analyse the overall change in the structure of the spectrum of industrial sectors' technological competencies in all component technological trajectories, the intra-technological distribution of all RTA_{ij} across all industrial sectors for all component technological trajectories at their point of takeoff is calculated. This is provided by the coefficient of variation (CV_{RTAij}) [represented by the standard deviation divided by the mean expressed as a percentage: $CV_{RTAij} = (\sigma_{RTAij}/\mu_{RTAij}) * 100\%$]. The CV_{RTAij} for each component technology is also listed in Tables 6.1 to 6.3. These results confirm the view that industrial sectors are coming together over time within the individual technological fields, provided by a falling CV_{RTAij}, on average, calculated across all industrial sectors. Especially, comparing the CV_{RTAij} for the same technological groups across several epochs of technological system takeoff confirms this evidence.

Moving beyond the technological systems the intra-industrial distribution (CV_{RTAij}) for each industrial sector across all technologies over time has also been calculated, and this evidence also indicates an overall fall historically, reflecting a trend towards an increase in the number of technologies specialised in by any given industrial sector. This also illustrates how industrial sectors have blurred and broadened their competencies.

6.5. CONCLUSION

In this chapter the co-evolution between innovation systems of technological trajectories and the complexes of industrial sectors making up the Chemical, Electrical / electronic, Mechanical and Transport 'industry systems' has been identified since the early interwar period. A competence bloc approach was applied as the selecting device, in order to establish the direction of those systems.

The chapter shows that industrial sectors are relatively more inclined to patent in the technological areas that are most closely associated with their industry, although this is less so over time. As demonstrated, a number of sub-sectors across industries have played an important role in several technological and economic upswings in the twentieth century. Patent records indicate that technological systems have become more complex over time, both with respect to the complexity of the component technological trajectories and the industrial sectors involved.

Overall, this chapter has mapped a historical trend across technological systems, in which industrial sectors appear to be coming together in adopting and developing the component technological trajectories of the systems. This is in turn also reflected in the industrial sectors' broadening of their technological competencies. This suggests an historical trend towards a more complex industrial base, and a more complex future development of technologies.

Hence, the results in this chapter support the notion of increased interrelatedness and complexity in technological evolution. That is, it is now commonly argued that technologies have become increasingly interrelated (see e.g. Chapter 3 of this book, Kodama 1986, 1992, Pavitt 1986, von Tunzelman 1995), and that the range of technological competencies required for success in any one innovation has increased (Patel and Pavitt 1994, Sharp 1989). The literature on collaborative alliances also uses this argument as one of the explanations for increased co-operation among firms, and between firms and other institutions (see Coombs *et al.* 1996 for an overview of this literature). Finally, it also demonstrates the view that knowledge bases used in technological development have become distributed across a wider spectrum of firms interacting in the socio-economic system (Andersen, Metcalfe and Tether 2000).

This chapter specifically illustrates how this increased technological interrelatedness is combined with a widening and increased complexity of corporate and industrial competencies, which (i) have a tendency to make industry boundaries less well-defined, (ii) change the nature of the competitive environment, (iii) which in turn have some feedback effect on subsequent innovations and hence on the technological landscape. It is therefore argued that, conceptually, an 'industry system' and its closest competitive environment can only be defined in non-static terms, as both the technological landscape in which it operates and its competencies change over time.

The competence bloc approach applied in the analysis of technological dynamics was very useful in illustrating how a technological system approach made a difference in the way we understand the boundaries of an industry. Firstly, if an industry needs to be understood in relation to its closest long-run dynamic competitive environment (c.f. innovative capacity) the standard industrial classification scheme (based upon product markets) simply provide us with very little insight, as firms' dynamic specialisations do not necessarily correspond to their counterpart product specialisation. Secondly, due to a changing innovative landscape underpinning the dynamics of industry, this also illustrates how we might reconsider the standard static view of industrial classification to become more flexible, if firms are to be grouped in relation to their closest competitors.

Applying a technological system approach (attempting to map the changing technological and industrial competitive landscape) was useful in capturing structures, detail (variety), historical changes, as well as overview. Prior to this chapter such studies have been generally of two kinds: one has been based on qualitative analysis by business and technology historians which have been rich in detail and historical evolution but lacked overview. The second are studies by economists which have contained much quantitative data but have often focused only on the short run or have been lacking in complementary qualitative support. Hence, one of the major advantages of the system perspective applied in this chapter is that it avoids the analytical tradeoff between complexity and overview, or between the long term and the detail, as it is able to provide both simultaneously.

Finally, the nature of the historical and comparative results suggest how systems develop and emerge, in which co-evolution is not only about catching up and accumulation, but also about new combinations of technologies and industrial structures, and the breaking down of old ones.

7. Types of Technological Competencies and Corporate Trajectories: The Variety of Firms and Path Dependency

Abstract

This chapter examines how firms within broad industries differ in their structure and type of technological competencies (specialisation), and how such differences matter for their technological profiles and their technological trajectories. A taxonomy of inter-company variety is developed. This classifies types of corporate technological competencies along three dimensions: the degree to which each firm exercises technological leadership within its industry; the extent to which the firm establishes competence in the development of the core technologies of its industry; and the overall breadth of technological specialisation of the firm.

The results illustrate how the spread of firms across the taxonomic groupings served to portray the variety of firms' technological profiles, even across firms with the same industrial base. Evidence also confirms how the intra-industry technological positions or profiles of firms are relatively stable and do not tend to fluctuate greatly over time but; when technological regimes change, firms' technological positions generally become eroded. This supports the view of the incremental, cumulative and path-dependent nature of the trajectories of corporate technological competencies. Overall, the results also suggest how the limits to the scope of the managerial discretion over corporate technological strategy is more complex than the one given by the specificity of an industry.

7.1. INTRODUCTION: INSIDE THE INDUSTRY

An interest has arisen within evolutionary economics as to why firms differ, and how it matters in relation to various theoretical issues, mainly in relation to the trajectories of firms. (Nelson 1991a, 1991b). Whereas neo-classical economic theory has little room for firm differences, such differences lie at the heart of enquiry for evolutionary economists and scholars of business management and strategy.

Firms are conventionally classified into industries based on product groups. Whereas such industrial classification based on product groups helps us to understand the context for innovation provided by product forms, uses and markets, the scale of production and vertical production linkages between firms (see Pavitt's *inter-industry* classification 1984b), this tells very little about the types of technological competencies used to develop and understand products and the linkages between them, or about the dynamic nature of corporate technological paths within industries and *intra-industry* competition between firms with alternative technological strategies. The latter is more related to the underlying technological capabilities of firms relative to their major competitors. In Chapter 6 it was even suggested that an alternative way to perceive an industry system could be in terms of the complex (or combination of) sectors of activity that exploit a common knowledge base, and are therefore involved in technological interchange regardless of the possible separation of their markets or lack of product-based associations.

While it has been demonstrated, and is now well known, that the variety of corporate technological strategies at a macro level are principally dependent upon the firm's industry (Pavitt 1984b), no-one has yet attempted to systematically identify and classify the variety of firm differences at a micro or intra-industry level. While there are limits to the scope of managerial discretion over corporate technological strategy given the specificity of industry paths (Patel and Pavitt 1998), within the limits of these constraints there is still room for different types of firms and different approaches to types of technological specialisation, and this chapter aims to examine these differences in terms of the evolution of alternative profiles of corporate technological specialisation found within industry groups historically.

Furthermore, some business scholars might accept the view that technological advance has been the key force driving economic growth over the past two centuries, with organisational change as its handmaiden. It is now commonly recognised that complex organisational structures are associated with the evolution of technology; but this chapter instead focuses on the dynamic technological capabilities of firms in relation to their corporate trajectories. Hence, whereas Nelson and Winter (1982) argued that technology is embodied in the organisational routines of firms, this study goes further and links the representation of the differentiated capabilities of firms to the composition of corporate technological competence. It is proposed here not only that firms differ with respect to their technological dynamic capabilities, as the localised nature of technological development creates a variety in the competence of firms, but that these differences can be categorised through a taxonomy of groupings of firms according to the characteristics of their corporate technological specialisation within their respective industries. Since changes in technological expertise and routines only occur gradually and within certain boundaries over time, it is believed that each firm becomes locked into some specific category of development path.[1]

Consequently, a classification of firms according to the characteristics of their types of technological competence, and the historical trajectories of firms within this classification, is worth investigating.

7.1.1. Chapter Outline

The first aim and procedure of this chapter is to establish a taxonomic classification of firms' technological profiles which is sufficiently general to be applicable to any industry. The construction of the taxonomy is based upon Andersen and Cantwell (1999), where the structure of firms' technological profiles was related to their capability to innovate, but this is not an objective in this chapter. The taxonomy will define the technological position of firms within their industries, in terms of three dimensions. Two of these dimensions relate directly to the actual composition of the profile of the firm's technological competence (the extent to which the firm is a technological leader within its broad industry group, and the extent to which the firm is involved in the development of the core technologies for its industry), and the third is the overall diversification of the firm's technological activity (i.e. the overall breadth of its technological competencies). The taxonomy is empirically grounded using measures of the technological position of the largest US and European firms developed from records of their US patenting activity from 1890 onwards. The composition of technological activity will be referred to in relation to each of the four broad industry groups; Chemicals, Electrical/electronics, Mechanical and Transport.

The relevance of the technological position of a firm is that (because technological paths are influenced by the inherited routines and expertise within each firm) it is likely to influence both the form and the future direction of the evolution of the spectrum of the firm's technological competence. In particular, the evolution of firms' technological competence as a path-dependent process will be analysed. The path-dependent processes in the trajectories of firms' relative position is investigated by calculating their underlying transition

[1] It has even been argued that the technological position of a firm within its industry (that is, the specific structure of its technological competence *vis-à-vis* others in the same industry) is likely to influence its current and future capability to innovate (Nelson and Winter 1982, Rosenberg 1982, 1994, Andersen and Cantwell 1999, Cantwell and Fai 1999.

probabilities of their historical behaviour with respect to the types and structure of their technological profiles. Finally, a short notion of the variety of firms is set out before reaching for a general conclusion

The analysis is considered in the context of technological epochs (Chapter 3) rather than technological systems (Chapter 6). The two broad historical epochs are 1920-40 (i.e. the interwar period) and 1960-90 (i.e. recent period), which were identified earlier in Chapter 3. The period in the middle - 1940-60 - will be cut out for two reasons: (i) the effects of the Second World War created fluctuations in corporate patenting activity so the results would be more a reflection of the war than anything else; and (ii) it was the period of a paradigm shift (as identified in Chapter 3).

If, alternatively, the unit of analysis was about the corporate trajectories of the individual firms participating in the technological systems identified in Chapter 6, we would also merely tell a story about the trajectories of those firms participating in developing and exploiting particular technologies in particular systems, and not shed light on trajectories of all firms in general. Besides, it would be impossible for technical reasons to consider corporate trajectories of firms' technological profiles at the technological system level (Chapter 6), since the total number of firms and number of technologies in such an approach change over time (this is a major point in the system framework), whereas we in this case of identifying the trajectories of individual firms historically need a long-run constant number of firms and technologies.

Applying a fixed intra-industry approach in this chapter, when identifying trajectories of corporate technological profiles historically, ought not to be seen in contrast to flexible system approach outlined in Chapter 6. In Chapter 6 it was also illustrated that technological trajectories within innovation systems tend to be patented by firms closest to the technological knowledge base, although it has been less so over time. Also, moving beyond the technological systems identified in Chapter 6, and if we instead consider total corporate patenting by the end of each epoch described above (i.e. in 1940 and 1990) of all 284 corporate groups selected for the development of this book (see Chapter 2), we can explicitly see from Tables 7.1 and 7.2 that, although firms are engaged in a broad range of patenting activities, chemical technologies still tend to be patented by chemical firms, electrical/electronic technologies tend to be patented by electrical/electronic firms, mechanical technologies by mechanical firms, and so forth. These results are even stronger historically, as we can see from Table 7.2. This, in accordance with the evidence in Chapter 6, also resonates that corporate technological competence in most technological fields is still primarily an industry-specific issue, although this relationship has been blurred over time.

Table 7.1. The total patent stock in 1990 based upon the 284 firms selected for this study, summed over all 399 technological fields, and organised by broad industrial group and broad technological group.

Broad technological group of patents	Broad industrial group of firms				
	Chemical	Electrical/ electronics	Mechanical	Transport	TOTAL
Chemical	79013.70	7927.77	8966.87	5282.10	101190.43
Electrical/ electronics	5391.07	49443.27	5187.77	8492.00	68514.10
Mechanical	26637.13	25427.27	26170.57	21738.53	99973.50
Transport	2270.77	945.30	2259.47	6696.53	12172.07
Other	1296.83	1887.00	891.43	665.77	4741.03
TOTAL	114609.50	85630.60	43476.10	42874.93	286591.13

Columns and rows may not sum exactly to the totals shown owing to rounding.

Table 7.2. The total patent stock in 1940 based upon the 284 firms selected from this study, summed over all 399 technological fields, and organised by broad industrial group and broad technological group.

Broad technological group of patents	Broad industrial group of firms				
	Chemical	Electrical/ Electronics	Mechanical	Transport	TOTAL
Chemical	15736.50	806.00	1470.53	849.77	18862.80
Electrical/ electronics	597.53	23040.50	1399.30	1218.63	26255.97
Mechanical	4502.43	9866.83	12841.35	6466.27	33677.07
Transport	406.67	633.30	765.40	1923.23	3728.60
Other	186.60	291.90	724.63	171.17	1374.30
TOTAL	21429.73	34638.53	17201.40	10629.07	83898.73

Columns and rows may not sum exactly to the totals shown owing to rounding.

Hence, if an investigation of firms' technological profiles in a historical context has to be based upon fixed classification, this is still best considered at the industry level. However, we can still analyse and compare the corporate trajectories across all industries.

7.1.2. Data Selection

In its taxonomic design, this chapter exploits some of the panel data characteristics that patent statistics can provide when classified by (i) type of technological activity at the very detailed 399 disaggregated class level, (ii) corporate ownership when classified at the 284 corporate group level, (iii) broad industry of firm, as well as (iv) over time. The organisation of the 399 technological sectors and 284 corporate groups are presented in Chapter 2. The analysis allows as well for corporate groups that operate under different names during different periods. For the purpose of this analysis, the technological and industrial sectors are sorted into their corresponding Chemical, Electrical/electronics, Mechanical and Transport broad groups (as identified in Chapter 2). Hence, Non-industrial technologies are left out, as there is no counterpart industrial group. In the design of empirical categories and methods, further restrictions have been imposed when selecting technological sectors and corporate patenting for the analysis. They are presented, where appropriately, in the text.

7.2. THEORETICAL FRAMEWORK: FROM TYPOLOGY TO TAXONOMY

The taxonomic classification of firms developed in this section defines the position of firms in technological space with reference to their patenting activity. The conceptual dimensions which underpin the classification scheme of firms' technological positions constitute a firm-typology (see section 7.2.1), while the firm-taxonomy (see section 7.2.3) represents the empirical application of the typology (developed in section 7.2.2.). This terminology is taken from Bailey (1994).

7.2.1. Conceptual Framework: The Firm Typology

Firms are classified according to their types of technological competence, where the profile of competence of each firm is measured by its pattern of technological specialisation, relative to its major rivals within its industry. Firms are classified within each industry by their technological position in terms of three dimensions. Two of these three dimensions - (i) the extent of the firm's technological leadership and (ii) the extent of the firm's participation in the core technologies of its industry - relate to the actual composition of the firm's technological competence and the other - (iii) the relative technological diversification of the firm - relates to the breadth of the firm's competence across technological.

(i) The first conceptual dimension considered is the extent of corporate technological leadership. A corporate technological leader is a firm which is specialised in the areas of the greatest technological opportunities, relative to other firms in the same industrial group.

Corporate technological leadership is measured by the extent to which the firm focuses on the *fastest growing* technological areas (i.e. areas of greatest technological opportunities) for its industry.

Hence, the notion of 'technological leadership' is only applied in relation to a certain type of corporate technological competence, and should not be confused with leadership in relation to market shares, or economic returns (which are important issues but not dealt with here). The latter might however be associated with the technological positions of firms.

(ii) The second conceptual dimension considered is the extent to which a firm focuses on the areas which represent the industry's core fields of technological development, relative to other firms in the same industrial group. The core sectors are those which were at the leading edge of development at some point in the past, and have now become a central line of activity in which most or all firms in the industry are engaged. A firm's competencies in the core fields of its industry is measured by the extent tc which the firm specialises in the technological areas with the *highest current level* of activity for its industry.

(iii) The third conceptual dimension considered is the diversification of the firm's competence across fields, relative to other firms in the same industrial group. The breadth, or diversification of the firm's competence across technological fields, is measured by the *overall dispersion* of the firm's technological activity within technologies for its industry. Besides, if a firm is narrowly specialised relative to other firms within its industry, at the same time as its competence base is outside the core fields which define its industry, it is regarded as being 'niche' focused.

The conceptual dimensions which underpin the classification scheme constitute a firm-typology is represented in Figures 7.1a. and 7.1b. The firm-taxonomy (see Figures 7.2a, 7.2b, and 7.3 in the next section) represents the empirical application of the typology.

Figure 7.1a. The typology

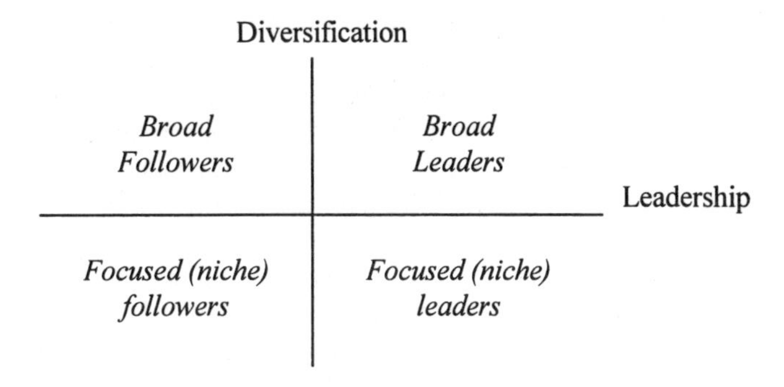

Figure 7.1b. The typology

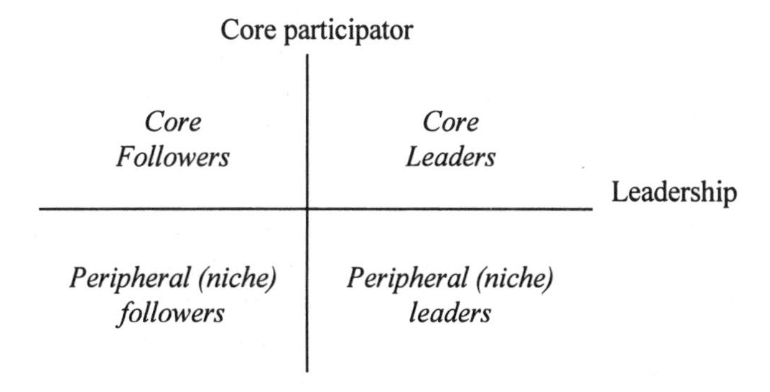

7.2.2. Empirical Applications of the Typology

(i) Areas of greatest technological opportunities and core technological sectors.

The conceptual exploration, and hence taxonomic quantification, defines the position of firms' technological competence with reference to their patenting activity in the fastest growing and core technologies of its industry as well as the overall size and scope or the breadth of its patenting activity. Hence, the fields of greatest technological opportunities and core fields of technological development have to be selected.

In order to avoid related problems that might be created by the inclusion of small patent classes, some extra restrictions on the selection of the fastest growing technological sectors have been imposed. Only sectors in which the total accumulated patent stock is at least 200, and in which the total accumulated corporate patenting is at least 50, at the end of each of two broad historical periods - the interwar, 1920-40, and the recent period, 1960-90 - have been selected. The two periods are chosen as they uniquely have been characterised by different periods of technological development governed by different technological epochs or paradigms (see Chapter 3), whereas the period of the war and its immediate aftermath is characterised by a structural break in corporate patenting activity and is therefore left out of the analysis.

The rate of growth of patenting (in order to select the areas of greatest technological opportunities) was then measured by the proportional change in accumulated patent stocks over the period in question.

From this evidence the fastest growing technological sectors over each period of technological development (1920-40 and 1960-90) were selected for the four broadly defined technological groups (that is, Chemicals, Electrical/electronics, Mechanical and Transport), by taking the 12 fastest growing patent classes in each period for Chemical, Electrical/electronic and Mechanical technologies. However, only six fast growing patent classes were selected for Transport technologies due to problems with small sectors, owing to the few sectors within that technological field. The fastest growing patent classes selected for each broad technological field and for both the interwar and recent period are set out in the Appendix (section 7.6.1), Tables 7.4-7.11.

The core technologies of a broad technological group (associated with a broad industrial group) which were at the leading edge of development at some point in the past and have now become a central line of activity, are defined as those technological sectors that have the largest absolute number of accumulated total corporate patents in the period in question, subject to one constraint. The restriction is that the sector's share of the patenting of the firms in an industry group is at least one and a half times that sector's share of the patenting activity of all firms. This restriction ensures that the core sectors represent the lines of activity that are especially relevant to each industry's development. Using these criteria, the core technological sectors selected for the interwar (1920-1940) and the recent (1960-90) period for the firms of each of the four broadly defined industry groupings were the twelve largest sectors in each period. However, only five core technological sectors were selected for Transport technologies in the interwar period in order to avoid the problems created by small numbers of patents. The selected core technologies are set out in the Appendix (section 7.6.2), Tables 7.12-7.19.

Note that no fast growing and core technological sectors within the broad Non-industrial technological group were selected, as there is no counterpart industrial group of firms (by definition).

(ii) Indices of the compositions of corporate technological competence.

For the purposes of an empirical application of that part of our typology that depends upon the structure of corporate technological competence, an index of corporate specialisation is constructed across *all* technological sectors (including fast growing as well as core technologies) for each broad industry. In this context, we examine in depth the cross-company values of the index of technological specialisation in the relevant fast growing and core sectors as described above; so, for example, when looking at chemical firms, we focus upon their specialisation relative to one another in the fastest growing chemical sectors and in the core technologies for the Chemical industry group.[2] Thus, corporate technological leadership, core participation and technological specialisation are intra-industry concepts, defined with respect to companies in a common industrial group.

Using the accumulated patent stock for each firm and for the total of all firms in the relevant broad industrial group, each firm's pattern of technological specialisation is then measured by an index termed revealed technological advantage (RTA). This measure was already introduced in Chapter 6 where it measured the specialisation of industrial sectors. However, in this case, the index is calculated for each firm relative to others within the same broad industrial group across the various sectors of technological activity. The RTA of a firm in a particular sector is given by the firm's share in that sector of US patents granted to companies in the same broad industrial group, relative to its overall share of all US patents granted to companies in the same broad industry.

Denoting by P_{ij} the accumulated number of patents granted to firm i within a particular industry in sector j; the RTA index for each firm in that industry is defined as follows:

(i) $$RTA_{ij} = \frac{P_{ij} / \sum_i P_{ij}}{\sum_j P_{ij} / \sum_{ij} P_{ij}}$$

The index varies around unity, such that values greater than unity suggest that a firm is positively or comparatively specialised in technological development in the sector in question, relative to its major international rivals. The RTA index has been calculated for all firms across all sectors of each broad industry for each year from 1920 to 1990.

The values of this index of technological specialisation in the fastest growing and in the core technological sectors which were each identified in the previous

[2] This approach is taken since firms within each industrial group are relatively more inclined to patent in the technological areas that are most associated with their industry (see 7.1. Introduction).

section are then selected. An unweighted average of the RTA measure over the relevant sectors for each firm i is calculated as a measure of the extent of the overall specialisation across both the fast growing and core technologies for its industry. It follows that where n is the total number of fast growing technologies selected (n being a subset of the total number of sectors, usually 12), firm i's average RTA in fast growing technologies, $RTAF_i$, is defined as follows:

$$\text{(ii)} \qquad RTAF_i = \sum_{f=1} RTA_{if} / n$$

The subscript f denotes an individual fast growing sector. As the index varies around unity, values greater than unity suggest that firm i is on average positively specialised in technological development in fast growing technologies, hence it can be termed a technological leader. Values lower than unity suggest that a firm is on average a technological follower.

Accordingly, if n is instead the total number of core technologies selected (which is also a subset of the total number of sectors, again usually 12) firm i's average RTA in core technologies, $RTAC_i$ is defined as follows:

$$\text{(iii)} \qquad RTAC_i = \sum_{c=1} RTA_{ic} / n$$

The subscript c denotes an individual core sector. Values greater then unity suggest that firm i is on average positively specialised in technological development in the core technologies, hence it is on average termed a core leader or core follower (depending on the RTAF value). Values lower than unity suggest that a firm is a peripheral leader or a peripheral follower.

Considerations

It here ought to be emphasised that the notion of 'leadership' (RTAF) or 'core participator' (RTAC) is defined with respect to the technological epoch (e.g. a leader is a firm with relative high competencies in the technological opportunities governed by the technological paradigm). On the other hand, those firms who are not 'leaders' or 'core participators' (i.e. not specialised in the areas of greatest technological opportunities or core fields, as governed by the paradigm) are by definition in this context defined as 'followers' or 'peripheral', although they may not even pursue a strategy of being specialised in the fast growing or core technologies, or catch up with the specialisation patterns of the leaders. With respect to our understanding of leadership, this also implies that the notion of being a 'follower' merely means a 'non-leader'.

A follower (i.e. non-leader of the paradigm), or a firm which is regarded as a technological 'peripheral' of a paradigm, may well hold greatest competencies in specific technologies outside the paradigm in question. All firms selected for this study are major technological active corporations at a global scale, as well as historically.

Although all the firms in this study are large (or quite quickly became large), some are larger than others, and these can be thought of as giant corporations by comparison with other relatively less giant or smaller firms in the investigation. However, as corporate technological 'leadership' or 'core participator' is defined here, smaller firms may be leaders or core participators if they are particularly oriented towards development in fastest growing or core areas of its industry, although the absolute level of their activity in these fields may be less than that of giant companies. Hence, giant corporations may be either technological leaders or followers or core participators or niche focused; what matters is neither their overall size nor their overall degree of specialisation, but rather the actual composition or profile of that specialisation - that is, whether their resources are especially geared towards development in the fastest growing or core fields.

Finally, as this taxonomy is based upon firms' technological profiles with respect to few (normally 12) selected patent classes (fast growing or of large size), the technological epochs is in this chapter are defined much narrowly than that of technological systems in Chapter 6 (which was based upon *S*-shaped growth cycles of much wider technological groups of related patent classes). Such a narrow technological epoch definition also implies that we built some narrowness and instability into the analysis of the trajectories of firms' technological profiles. This has implications on their technological positions within the taxonomy, and might also have implications on the degree to which they move in a path dependent fashion over longer periods of time.

(iii) Index of the breadth of the firms' competence

Nevertheless, the absolute size of activity does have an impact on the cross-sectoral distribution of a company's activity, as it tends to regulate the firm's overall degree of technological specialisation or diversity - that is, larger firms tend to be more technologically diverse (as suggested, for example, by Chandler 1990). This relationship has also more recently been confirmed by Cantwell and colleagues[3] using the same US database on which this book is based. Hence, larger firms cover a wider range of technological development, and so typically have a more broadly based spectrum of technological competence. Therefore,

---

[3] Cantwell and Bachmann (1998), Cantwell and Fai (1999) and Cantwell and Santangelo (2000)

we need to distinguish firms by their (relative) breadth of their competence, as well as by whether they are technological leaders or followers, and by the extent of their participation in core sectors.

Hence, the third dimension in this technological classification of firms takes into consideration the cross-firm diversification distribution of the absolute level of technological competence. In taxonomic terms a measure of diversification (DIV) is achieved by using the inverse of the concentration index, the co-efficient of variation (CV), where the CV [represented by the standard deviation divided by the mean expressed as a percentage: $CV_{RTAij} = (\sigma_{RTAij}/\mu_{RTAij}) * 100\%$] is calculated for each firm's RTA value across all technological sectors for its industry.

(iv) $$DIV_{ij} = 1/ CV_{(RTAij)}$$

The DIV_{ij} index measures breath of the total patenting of firm i across all sectors j for its industry. In this context a high value of the DIV measure reflects a high degree of diversification on average (or less degree of concentration). That is the firm is broadly focused. On the contrary, a small value of DIV measure reflects a smaller degree of diversification on average (or higher degree of concentration). i.e. the firm is more specialised.

In this context, if a firm i is narrowly specialised relative to other firms within its industry (i.e. DIV_i < average), at the same time as its competence base is outside the core fields which define its industry (i.e. $RTAC_i < 1$), it is regarded as being 'niche' focused.

7.2.3. Taxonomic Classification of the Technological Profile of Firms

Figure 7.2a. The taxonomy

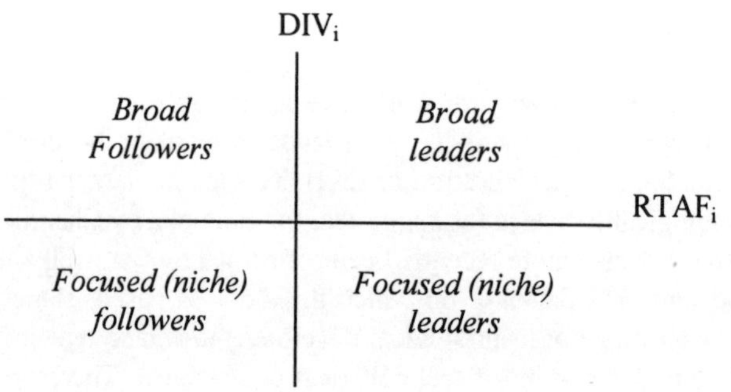

	DIV_i	
Broad Followers	*Broad leaders*	
		$RTAF_i$
Focused (niche) followers	*Focused (niche) leaders*	

Figure 7.2b. The taxonomy

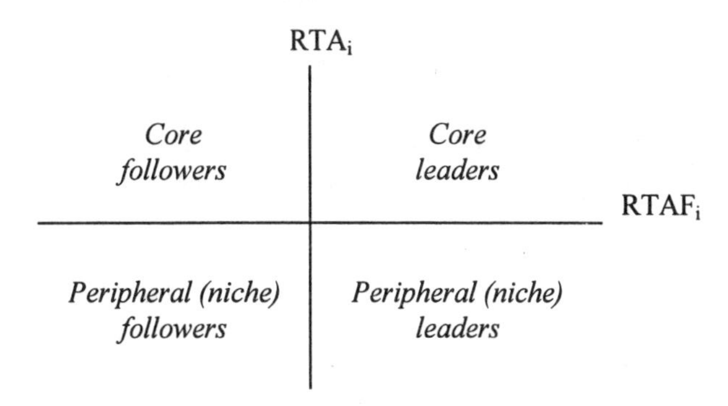

For the purpose of classifying firms within each broad industry group, the typology presented graphically in Figures 7.1a and 7.1b can now be empirically transformed into a taxonomy as illustrated in Figures 7.2a and 7.2b, and further converted into tabular form, where the firms are classified into eight groups as portrayed in Figure 7.3.

Figure 7.3. Taxonomic classification of the position of firms' technological profiles

	Followers: $RTAF_i < 1$		Leaders: $RTAF_I > 1$	
	Outside core fields: $RTAC_i < 1$	Inside core fields: $RTAC_i > 1$	Outside core fields: $RTAC_i < 1$	Inside core fields: $RTAC_i > 1$
More narrowly specialised: $DIV_i <$ ave	**1** Niche peripheral followers	**2** Focused core followers	**3** Niche peripheral leaders	**4** Focused core leaders
Broader specialised: $DIV_i >$ ave	**5** Broad peripheral followers	**6** Broad core followers	**7** Broad peripheral leaders	**8** Broad core leaders

Note:

1: Niche (often small) firms which are not comparatively specialised in either core or fast growing technologies for their industry; hence they act as niche peripheral followers.

2: Focused (often small) firms which are comparatively specialised in core technologies but are not specialised in fast growing technologies for their industry; hence they act as focused core followers.

3: Niche (often small) firms which are not comparatively specialised in core technologies but are specialised in fast growing technologies for their industry; hence they act as niche peripheral leaders.

4: Focused (often small) firms are comparatively specialised in both core and fast growing technologies for their industry; hence they act as a focused core leaders.

5: Diversified (often large) firms which are not comparatively specialised in either core or fast growing technologies for their industry; hence they act as broad peripheral followers.

6: Diversified (often large) firms which are comparatively specialised in core technologies but are not specialised in fast growing technologies for their industry; hence they act as broad core followers.

7: Diversified (often large) firms which are not comparatively specialised in core technologies but are specialised in fast growing technologies for their industry; hence they act as broad peripheral leaders.

8: Diversified (often large) firms which are comparatively specialised in both core and fast growing technologies for their industry; hence they act as broad core leaders.

7.3. THE PATH-DEPENDENT TRAJECTORIES OF FIRMS' TECHNOLOGICAL PROFILES

However, as stated in the Introduction to this Chapter, what is interesting is not primarily the general structure of the position of firms, but an examination of how the underlying trajectories of the firms' technological position evolves over time. Since the evolution of the firm's profile of technological competence is expected to be a path-dependent process (see Introduction, section 7.1), it is relevant to examine whether there is a tendency for firms to move only within certain restricted regions of the taxonomy as illustrated in Figure 7.3, depending upon their particular historical pattern of technological development, size, and profile of accumulated technological competence.

When analysing the paths of firms within certain restricted regions of the taxonomy, as illustrated in Figure 7.3, help can be drawn from probability theory. A stochastic process in which a discrete random variable, say firm i at time t - i.e. i(t) - may change state (i.e. value, or taxonomic grouping in this case) at times t_n, t_{n+1}, t_{n+2}, . . . (usually equally spaced) is called a Markov chain, if the conditional distribution of $i(t_{n+1})$ depends only on $i(t_n)$ and not on the value of i at any earlier time. This is often expressed by saying that the state of the system in the future is dependent on the present stage, but is unaffected by its history.

In this particular case, each firm i can be positioned within one of eight different stages (or categories as identified in Figure 7.3) within its industry, and it may change position over time. At any transition in time, t_n, there are eight probabilities, $p11, p12, p13, p14, p15, p16, p17, p18$, that the system, if in stage 1, will stay in 1, or change to stage 2, 3, 4, 5, 6, 7 or 8. Similarly there are eight probabilities of transition, if the firm is grouped in stage 2, 3, 4, 5, 6, 7, or 8 at t_n.

In this context $p11$ for time t_{n+1} is calculated as the number of firms, i, staying in category 1, divided by the total number of firms, i, grouped in category 1 in the previous period, t_n. $p12$ for time t_{n+1} is calculated as the number of firms moving into category 2, divided by the total number of firms, i, grouped in category 1 in the previous period, t_n. $p13$ for time t_{n+1} is calculated as the number of firms moving into category 3, divided by the total number of firms, i, grouped in category 1 in the previous period, t_n and so forth. In such a way each of the rows of probabilities in the transition matrix (see equation (v)) adds up to 1 (e.g. $p11 + p12 + p13 + p14 + p15 + p16 + p17 + p18 = 1$)

The matrix of the transition probabilities is called a transition matrix and is here denoted P:

$$(v) \qquad P = \begin{pmatrix} p11 & p12 & p13 & p14 & p15 & p16 & p17 & p18 \\ p21 & p22 & p23 & p24 & p25 & p26 & p27 & p28 \\ p31 & p32 & p33 & p34 & p35 & p36 & p37 & p38 \\ p41 & p42 & p43 & p44 & p45 & p46 & p47 & p48 \\ p51 & p52 & p53 & p54 & p55 & p56 & p57 & p58 \\ p61 & p62 & p63 & p64 & p65 & p66 & p67 & p68 \\ p71 & p72 & p73 & p74 & p75 & p76 & p77 & p78 \\ p81 & p82 & p83 & p84 & p85 & p86 & p87 & p88 \end{pmatrix}$$

If $p11 = p22 = p33 = p44 = p55 = p55 = p66 = p77 = p88 = 1$, then the system never changes stage, and the trajectories of corporate competencies may be considered as being totally persistent or path-dependent.

If the Markov processes are considered over several periods, say t_{n+1}, t_{n+2}, t_{n+3}, $p11$ is calculated as the sum of firms, i, staying in category 1 for each transition periods t_{n+1}, t_{n+2}, and t_{n+3}, divided by the sum of firms, i, grouped in category 1 in the corresponding previous periods, t_n, t_{n+1}, and t_{n+2}. $p12$ is calculated as the sum of firms, i, moving into category 2 for each transition periods, t_{n+1}, t_{n+2}, and t_{n+3}, divided by the sum of firms, i, grouped in category 1 in the corresponding previous periods, t_n, t_{n+1}, and t_{n+2}.

When investigating the evolution of firms' trajectories while testing for the degree to which the behaviour of firms is dependent upon certain paths, all firms which have existed throughout the period 1930 to 1990 (i.e. patents > 0) have been taken into the analysis. We have allowed as well for groups that operated under different names during the period. The analysis is then divided into the two historical technological epochs, 1920-1940 and 1960-1990 mentioned in the Introduction (Section 7.1). However, to avoid problems created by small numbers in patenting activity (which creates only random RTA results) only firms with significant large accumulated patent stock in 1930 and 1940 (i.e. patent stock > 50 in both 1930 and 1940) as well as in 1960-90 (i.e. patent stock > 50 in 1960, 1970, 1980 and 1990) has been extracted. The analysis of corporate trajectories in the interwar period (where the technologies of great opportunities and core importance are calculated over the period 1920-1940) can, at earliest, be investigated from 1930, as any point previous to 1930 would reduce the number of firms significantly.

Figure 7.4. Transition matrix 1930-1940

P11	P12	P13	P14	P15	P16	P17	P18
0.53	0.12	0.23	0	0.06	0	0.06	0.06
P21	P22	P23	P24	P25	P26	P27	P28
0.25	0.33	0	0	0.08	0.33	0	0
P31	P32	P33	P34	P35	P36	P37	P38
0.4	0	0.2	0.2	0	0	0.2	0
P41	P42	P43	P44	P45	P46	P47	P48
0	0.14	0.14	0.43	0	0	0.14	0.14
P51	P52	P53	P54	P55	P56	P57	P58
0.09	0	0	0	0.54	0.09	0.27	0
P61	P62	P63	P64	P65	P66	P67	P68
0	0.09	0	0	0.09	0.64	0.18	0
P71	P72	P73	P74	P75	P76	P77	P78
0	0	0.33	0	0	0	0.67	0
P81	P82	P83	P84	P85	P86	P87	P88
0	0.33	0	0	0	0.33	0	0.33

The probabilities values within each row do not always add up to 1, due to rounding.

This leaves us with 69 firms in total for the interwar period, consisting of 20 Chemical firms, 13 Electrical/electronic firms, 23 Mechanical and 13 Transport firms. For the recent period 118 firms enter the analysis, distributed across 38 Chemical firms, 18 Electrical/electronic firms, 40 Mechanical and 22 Transport firms.

When pooling all the transition matrix, P, data (based upon a taxonomy developed for each industry individually, for the reasons set out in the Introduction, section 7.1), the underlying transition probabilities for all industries together are, with respect to the interwar period, 1930-1940 shown in Figure 7.4, and with respect to the recent transition periods 1960-1970-1980-1990 shown in Figure 7.5.

Figure 7.5. Transition matrix 1960-1970-1980-1990

P11 0.72	P12 0.13	P13 0.10	P14 0	P15 0.04	P16 0.01	P17 0	P18 0
P21 0.16	P22 0.58	P23 0.03	P24 0.05	P25 0.03	P26 0.09	P27 0.02	P28 0.03
P31 0.19	P32 0.02	P33 0.59	P34 0	P35 0.07	P36 0	P37 0.11	P38 0
P41 0.05	P42 0.10	P43 0.10	P44 0.60	P45 0	P46 0.05	P47 0.05	P48 0.05
P51 0.10	P52 0.02	P53 0.02	P54 0	P55 0.60	P56 0.13	P57 0.10	P58 0.02
P61 0	P62 0.11	P63 0	P64 0.01	P65 0.06	P66 0.70	P67 0	P68 0.11
P71 0	P72 0	P73 0.21	P74 0	P75 0.08	P76 0.08	P77 0.58	P78 0.04
P81 0.04	P82 0	P83 0.04	P84 0.04	P85 0	P86 0.29	P87 0	P88 0.58

The probabilities values within each row do not always add up to 1, due to rounding.

Those transitions which are most likely (or are as likely as those which are most likely) are illustrated in bold frames. The results show how firms, on average, are more likely *not* to change their types of technological competencies or corporate technological strategies, than to move into other types of competence profiles. This evidence is very strong in recent period.

That persistency, or path dependency of firms, is less strong in the interwar period could be due to several reasons. One reason could be that, as the interwar period was a period in which broad technological groups' trajectories of opportunities diverged (as identified in Chapter 3), this was shaking up firm and intra-industry strategies with respect to the types of technological competencies they wished to pursue. It could also be a period in which industry structures were not yet formed (as opposed to the more mature manufacturing industries today). Especially, in Chapter 3 it was suggested how the mobility in the science based fields within the interwar period, may reflect how relatively newly established industries were going through some degree of 'search and selection' processes of development. Finally, the results for the interwar period could also just stress that we are dealing with fewer observations, so that the estimated probability values become more random. Not just are there fewer firms eligible for the study (due to the restrictions imposed upon the data), but the underlying transition matrix is also calculated over only one transition, as opposed to three in the more recent period.[4] Most likely the results reflect a combination of all those factors.

Investigating the transition values at the industry level, such an exercise is pointless for the interwar periods due to problems with small numbers. However, for the recent period (1960-1990) the transition matrixes at the industry level also show a high degree of path dependency, except for only 4 taxonomic groupings in total across all industries. This applies to taxonomic groupings 6 and 7 in the case of the Chemical industry, taxonomic grouping 3 in the case of the Electrical industry, and taxonomic grouping 8 in the case of the Transport industry. That firms in these few taxonomic groupings are more likely to change their type of corporate technological competence is presumably a small number problem, as 3 of those groupings only have 2 or 3 observations in total entering the analysis during all the periods of transition, and one grouping got only 7 observations in total.

As a matter of comparison it is interesting to investigate how the position of firms is likely to change as technological epochs or regimes change (i.e. when new technological opportunities and new core sectors occur). For this purpose a transition matrix (see Figure 7.6) is calculated across the position of firms by the end of the interwar technological epoch, i.e. 1940, to the beginning of the recent technological epoch, i.e. 1960. However, 1950 is left out of the analysis as the patenting activity (and hence the position) of firms is mainly a reflection of the war (high fluctuations occur on firms patenting behaviour within the period up to and shortly after 1950). Using the period 1940-1960, as opposed the period 1940-1950-1960, decreases the number of observations which otherwise could have been used for this type of analysis. However, by pulling all the industries again, such problems with small numbers should be minimised.

The cross-technological-epoch transition matrix is calculated only for the firms who were eligible for the estimations of the transition matrix in *both* the interwar technological epoch (1930-1940) and the recent epoch (1960-1990).

---

[4] Obviously the probability numbers calculated for recent period (Figure 7.5) is based upon more observations, as this is calculated for more firms and across three period of transition, $t_{1960\text{-}1970}$, $t_{1970\text{-}1980}$ and $t_{1980\text{-}1990}$; as opposed to fewer firms and only one periods of transition, $t_{1930\text{-}1940}$, in the interwar period (Figure 7.4).

Hence, the firms of the interwar technological epoch which were not eligible for calculation of the transition matrix within the new technological epoch is left out of the analysis. This leaves us with 62 firms in total.

Figure 7.6. Transition matrix 1940-1960 (for selected firms*)

P11 0.5	P12 0.17	P13 0.08	P14 0.08	P15 0	P16 0	P17 0.08	P18 0.08
P21 0.43	P22 0.29	P23 0	P24 0.14	P25 0	P26 0.14	P27 0	P28 0
P31 0.14	P32 0.14	P33 0.14	P34 0.14	P35 0.14	P36 0.29	P37 0	P38 0
P41 0.25	P42 0.25	P43 0.25	P44 0	P45 0	P46 0.25	P47 0	P48 0
P51 0	P52 0.25	P53 0	P54 0	P55 0.12	P56 0.50	P57 0.12	P58 0
P61 0.08	P62 0.08	P63 0	P64 0	P65 0.25	P66 0.50	P67 0	P68 0.08
P71 0.10	P72 0	P73 0.10	P74 0.10	P75 0.20	P76 0.30	P77 0.10	P78 0.10
P81 0	P82 0	P83 0	P84 0	P85 0	P86 0.50	P87 0.50	P88 0

The probabilities values within each row do not always add up to 1, due to rounding.
** Only those 'eligible' firms who participated in both the interwar system (1930-1940) and recent system (1960-1990) are use for the analysis.*

It is not surprising to see that only very little path dependency in the position of firms occur, due to the change in technological regime (with new fast growing and new core technologies). Only firms grouped within taxonomic grouping 1 and 6 have highest probability to stay within their taxonomic position. Also, although the firms in taxonomic grouping 4 move widely, it is mostly within the path of being narrowly specialised. However, it is interesting (but not surprising) to see how most firms participating in the interwar technological epoch have the highest probability to move into taxonomic grouping 6 within the new system, which is broad core followers. Hence, almost no-matter what position firms had in the interwar system, if they survived to participate in the new system with at least the same patenting insensitivity, they appeared to be followers, broad specialised and in core technologies of its industry. This may be what was fast growing in the past (i.e. areas of great technological opportunities) has now grown to become of core importance of the industry. This especially applies to the Chemical and Transport broad technological groups, but also applies to some extent to the Electrical/ electronics and Mechanical broad technological groups (see Tables 7.4-7.19 in the Appendix, section 7.6). Hence, firms who in the past developed a strategy of becoming specialised in, or catching up with, technological fields of new opportunities, now find themselves specialising in the areas which have become of core importance for their industry.

Table 7.3. The absolute distribution of firms in accordance with their degree of leadership- or followership-persistency within and across technological epochs

INTERWAR technological epoch: 1930-1940:	Number of firms
Leaders throughout the interwar epoch.	13
Followers throughout the interwar epoch.	41
Rest of interwar firms.	15
TOTAL interwar firms.	69
RECENT technological epoch: 1960-1970-1980-1990:	Number of firms
Leaders throughout the recent epoch.	15
Followers throughout the recent epoch.	57
Rest of recent firms	36
TOTAL recent firms.	118
TWENTIETH CENTURY technological epochs: interwar (1930-1940) and recent period (1960-1970-1980-1990):	Number of firms
Leaders throughout the twentieth century's technological epochs.	3
Followers throughout the twentieth century's technological epochs	22
Rest of firms.	37
TOTAL eligible historical firms participating throughout the twentieth century's technological epochs.	62

That more firms exercise a continuity of position within a technological epoch, rather than across epochs (between which a structural break occurs in the pattern of technological opportunities and sectors of core importance) indicates that it is not easy for firms to develop a competence in new unrelated areas. These results are suggestive of the incremental, cumulative and path-dependent nature of the development of types of corporate technological competencies within technological epochs governed by shifting paradigms. In line with the findings in the case of the evolution of leading firms across technological epochs, we also see in Chapter 8 that the technological competencies of firms within a technology change very little historically. That is, firms tend to keep their types of corporate capabilities (e.g. degree of leadership) within a technology historically, but when technological epochs or regimes change, firms' technological position, as identified within an old technological regime, tend to become eroded.

For example among the 62 historical firms (i.e. those which participated in both systems) only 3 firms remained as leaders (i.e. stayed with the classification of 3, 4, 7, or 8) throughout both the interwar and recent periods (i.e. 1930, 1940, leaving out 1950, 1960, 1970, 1980 and 1990). See Table 7.3.

7.4. THE VARIETY OF FIRMS

The identification and trajectories of the firms' technological profiles sorted by their positions by the end of each period (i.e. 1940 and 1990) is illustrated in the Appendix (section 7.6.3). This includes 69 firms for the interwar period and 118 firms for the recent period. As mentioned in section 7.3, only firms which are large enough in 1930 and 1940 (i.e. patents > 50 in both 1930 and 1940) or in 1960-90 (i.e. patents > 50 in all of 1960, 1970, 1980 and 1990) have been extracted for the analysis of Markov chains, although the calculation of degree of leadership (RTAF), core-participation (RTAC) and technological diversification (DIV) in section 7.2 is done across *all* firms which patented in the US historically.

From the overall distribution of firms across the taxonomic groupings (Appendix, Tables 7.20-7.27), we see how there is no, or very little, tendency for firms to cluster in certain of the 8 specific categories of the taxonomy. This finding applies across all industries over the entire period of the analysis, although to a lesser extent for the interwar period, when the distribution of firms tends to be more polarised (owing to the fewer observations). Nonetheless, the bulk of firms tend to be technological followers, falling into categories 1, 2, 5 and 6 of the taxonomy (Figure 7.3).

The spread of firms across the taxonomic groupings serve to illustrate the variety of firms, and help to confirm the usefulness of the typology in practice, and it also illustrates how none of the three dimensions (leadership, core participation and diversification) is particularly correlated across firms with either of the other two. (That the variables or indexes of firms' types of corporate technological competencies developed here, with respect to leadership, core participation and diversification, are not correlated with either of the other two, was also demonstrated in Andersen and Cantwell (1999) which examined the correlation between those independent variables (i.e. tested for multi-correlation), when using them as regressors in another study on firms' capability to innovate.)

Overall, the results also demonstrate how it is wrong to perceive an industry at the aggregate level as different business models apply.

As an illustration of the profiles of firms in relation to the taxonomy in question, we can view a few examples of the technological positions of firms historically. Courtaulds is a textiles firm, so obviously it is highly technologically specialised by the standards of the chemical firms with which it is grouped (i.e. low diversification) and outside the core chemical sectors. While it may have been a leader at the time when rayon was invented at the beginning of this century, it quickly fell behind in the growth of corporate research from the interwar years onwards. So its pattern - region 1 throughout - isn't surprising. Following the merger that created ICI in 1926, the firm moved into a wider range of chemical research, and it has strived to become a chemical generalist (moving into regions 5 or 6). Otherwise been well represented in the core sectors of chemicals (region 2), it has never managed to establish itself as a leader of the technological paradigms in questions. The case of Standard Oil New Jersey (later Exxon) may tell you as much about the evolution of chemical technology as it does about the firm's own trajectory. It was a leader in the interwar period when the focus of development included great interest in oil-related technologies, although it moved out of the core fields of its industry (moved from region 4 to 7) just before the technological regime of the industry went through structural change. After the war chemical advances were mainly in other areas, but petrochemicals remained central and Exxon was positioned in region 6 of the taxonomy as a broad core follower. However, more recently Exxon has established links between mineral and biotech research, and regained leadership within the Chemical industry. It has even managed to become a broad core leader (i.e. positioned within region 8) of the technological paradigm in questions. IBM shows a high degree of corporate evolution. It started out as a mechanical typewriter firm. Hence, its mechanical technologies were neither fast growing nor core, and the firm was non-diversified by electrical industry standards - hence it was positioned in region 1 in the interwar period.). However, it moved to being an office computing equipment specialist with great activity in both fast growing and now core fields of its industry. Although it was at first not very diversified (positioned in region 4), it subsequently has become a computer equipment generalist (region 8). General Electric (GE, not the British GEC) began as a traditional electrical generalist (in illumination) with a good electrical engineering base. Thus, it was highly diversified and core, but not focused in the fastest growing developments (region 6). However, after the war it used its all-round strength to gravitate towards the faster growing activities while retaining too much general electrical engineering to be specialised in the core areas of the electrical industry (moving between regions 5 and 7 of the taxonomy).

The aim of this chapter is not business historical, but the usefulness of the taxonomy as a point of reference in understanding the evolution of the technological profiles of firms should be evident. (See Appendix, (section 7.6.3, Tables 7.20-7.27, for further information concerning the evolution of corporate technological profiles)

Classifying firms at micro or intra-industry level also contribute to Patel and Pavitt (1998) who argued how there are limits to the scope of managerial discretion over corporate technological strategy given the specificity of industry paths. The analysis in this chapter demonstrates that even within the limits of these constraints there is still room for different types of firms and different approaches to types of technological specialisation and profile, and that this has implications for firms' subsequent evolution paths.

7.5. CONCLUSION

With reference to some of the principal determinants of corporate competitiveness a firm taxonomy has been developed. The firm taxonomy was then subsequently used to examine how the technological position of firms influenced their technological trajectories.

It was illustrated how the intra-industry technological positions of firms are relatively stable and do not tend to fluctuate greatly over time, but when technological regimes change, firms' relative technological positions generally become eroded. These results are suggestive of the incremental, cumulative and path-dependent nature of the trajectories of corporate technological competence.

It was also illustrated how the spread of firms across the taxonomic groupings served to portray the variety of firms' technological profiles, even across firms with the same industrial base. This also indicates how different business models apply across firms within industry. Hence, the limits to the scope of the managerial discretion over corporate technological strategy is more complex than the one given by the specificity of an industry.

The results evidently establish the usefulness of an intra-industry taxonomy of types of corporate technological competencies in practice.

7.6. APPENDIX

7.6.1. The Fastest Growing Technological Sectors

Table 7.4. The fastest growing Chemical technological sectors 1920-1940

Technological Group	Patent Class / Subclass: Sector of Activity
5 Chemical processes	Class 62: Refrigeration
9 Synthetic resins and fibres	Classes 524, 525, 526, 528: Parts of class series on synthetic resins
11 Other organic compounds	Classes 536, 558, 568, 570: Parts of the class series on organic compounds
12 Pharmaceuticals and biotechnology	Class 514: Drug, bio-affecting and body treating compositions
51 Coal and petroleum products	Class 208: Mineral oils; processes and products
	Class 585: Chemistry, hydrocarbons

Table 7.5. The fastest growing Chemical technological sectors 1960-1990

Technological Group	Patent Class / Subclass: Sector of Activity
4 Agricultural chemicals	Class 71: Chemistry, fertilizers, pesticides
5 Chemical processes	Class 210: Liquid purification or separation
6 Photographic chemistry	Class 430: Radiation imagery chemistry-processes, compositions or products
7 Cleaning agents and other compositions	Class 512: Perfume compositions
9 Synthetic resins and fibres	Classes 521, 522, 523: Parts of class series on synthetic resins
11 Other organic compounds	Class 540: Part of class series on organic compounds
12 Pharmaceuticals and biotechnology	Classes 424, 514: Drug, bio-affecting and body treating compositions
	Class 435: Chemistry: molecular biology and microbiology
	Class 436: Chemistry: analytical and immunological testing

Table 7.6. The fastest growing Electrical/electronic technological sectors 1920-1940

Technological Group	Patent Class / Subclass: Sector of Activity
33 Telecommunications	Class 332: Modulators
	Class 370: Multiplex communications
	Class 455: Telecommunications
35 Special radio systems	Class 342: Communications, directive radio wave system and devices
	Class 343: Communications, radio wave antennas
36 Image and sound equipment	Class 358: Pictorial communication; television
38 Electrical devices and systems	Class 328: Miscellaneous electron space discharge device
	Class 330: Amplifiers
	Class 331: Oscillators
	Class 333: Wave transmission lines and networks
39 Other general electrical equipment	Class 62: Refrigeration
40 Semiconductors	Class 357: Active solid state devices, e.g. Transistors, solid state diodes

Table 7.7. The fastest growing Electrical/electronic technological sectors 1960-1990

Technological Group	Patent Class / Subclass: Sector of Activity
33 Telecommunications	Class 375: Pulse or digital communications
34 Other electrical communication systems	Class 341: Coded data generation or conversion
	Class 382: Image analysis
38 Electrical devices and systems	Class 372: Coherent light generators
39 Other general electrical equipment	Class 437: Semiconductor device manufacturing: process
40 Semiconductors	Class 307: Electrical transmission or interconnection systems
	Class 357: Active solid state devices, eg. transistors, solid state diodes
41 Office equipment and data processing systems	Class 360: Dynamic magnetic information storage and retrieval
	Class 364: Electrical computers and data procession systems
	Class 365: Static information and storage and retrieval
	Class 371: Error detection/correction and fault detection/recovery
52 Photographic equipment	Class 355: Photocopying

Table 7.8. The fastest growing Mechanical technological sectors 1920-1940

Technological Group	Patent Class / Subclass: Sector of Activity
1 Food and tobacco products	Class 426: Food or edible material: processes, compositions and products
13 Metallurgical processes	Class 29: Metal working
	Class 148: Metal treatment
14 Miscellaneous metal products	Class 206: Special receptacle or package
18 Paper making apparatus	Class 162: Paper making and fiber preparation
29 Other general industrial equipment	Class 165: Heat exchange
50 Non-metallic mineral products	Class 65: Glass manufacturing
	Class 428: Stock material or miscellaneous articles
	Class 501: Compositions: ceramic
53 Other instruments and controls	Class 350: Optics, systems and elements
	Class 352: Optics, motion pictures
	Class 356: Optics, measuring and testing

Table 7.9. The fastest growing Mechanical technological sectors 1960-1990

Technological Group	Patent Class / Subclass: Sector of Activity
13 Metallurgical processes	Class 228: Metal fusion bonding
14 Miscellaneous metal products	Class 623: Prosthesis (artificial body members), parts thereof or aids and accessories therefor
16 Chemical and allied equipment	Class 502: Catalyst, solid sorbent, or support thereof, products and process
	Class 503: Record receiver having plural leaves or a colourless colour former, method of use or developer therefor
50 Non-metallic mineral products	Class 65: Glass manufacturing
	Class 428: Stock material or miscellaneous articles
53 Other instruments and controls	Classes 128, 604: Part of class series on Surgery
	Class 250: Radiant energy
	Class 346: Recorders
	Class 350: Optics, systems and elements
	Class 356: Optics, measuring and testing

Table 7.10. The fastest growing Transport technological sectors 1920-1940

Technological Group	Patent Class / Subclass: Sector of Activity
42 Internal combustion engines	Class 123: Internal combustion engines
43 Motor vehicles	Class 180: Motor vehicles
	Class 296: Land vehicles, bodies and tops
44 Aircraft	Class 244: Aeronautics
47 Other transport equipment	Class 280: Land vehicles
49 Rubber and plastic products	Class 264: Plastic and nonmetallic article shaping or treating process

Table 7.11. The fastest growing Transport technological sectors 1960-1990

Technological Group	Patent Class / Subclass: Sector of Activity
42 Internal combustion engines	Class 123: Internal combustion engines
43 Motor vehicles	Class 180: Motor vehicles
	Class 296: Land vehicles, bodies and tops
45 Ships and marine propulsion	Class 114: Ships
47 Other transport equipment	Class 293: Vehicle fenders
49 Rubber and plastic products	Class 264: Plastic and nonmetallic article shaping or treating process

7.6.2. The Core Technological Sectors

Table 7.12. The core Chemical technological sectors 1920-40

Technological Group	Patent Class / Subclass: Sector of Activity
3 Inorganic chemicals	Class 423: Chemistry, inorganic
7 Cleaning agents and other compositions	Class 252: Compositions
	Class 106: Compositions, coating or plastic
10 Bleaching and dyeing	Class 8: Bleaching and dyeing of textiles and fibres
11 Other organic compounds	Classes 534, 568, 552, 562, 564, 548: Parts of the class series on organic compounds
50 Non-metallic mineral products	Class 428: Stock material or miscellaneous articles
51 Coal and petroleum products	Class 208: Mineral oils; processes and products

Table 7.13. The core Chemical technological sectors 1960-90

Technological Group	Patent Class / Subclass: Sector of Activity
3 Inorganic chemicals	Class 423: Chemistry, inorganic
4 Agricultural chemicals	Class 71: Chemistry, fertilizers
7 Cleaning agents and other compositions	Class 252: Compositions
9 Synthetic resins and fibres	Classes 524, 525, 528: Parts of the class series on synthetic resins
11 Other organic compounds	Class 568, 548, 544: Parts of the class series on organic compounds
12 Pharmaceuticlas and biotechnology	Class 514: Drug, bio-affecting and body treating compositions
16 Chemical and allied equipment	Class 502: Catalyst, solid sorbent, or support thereof, products or process
51 Coal and petroleum products	Class 208: Mineral oils; processes and products

Table 7.14. The core Electrical/electronic technological sectors 1920-40

Technological Group	Patent Class / Subclass: Sector of Activity
29 Other general industrial equipment	Class 303: Fluid pressure brake and analogous systems
33 Telecommunications	Class 379: Telephonic communication
	Class 455: Telecommunications
	Class 178: Telegraph
37 Illumination devices	Class 313: Electric lamp and discharge devices
	Class 315: Electrical lamp and discharge devices, systems
38 Electrical devices and systems	Class 361: Electricity, electrical systems and devices
	Class 200: Electricity, circuit marker and breakers
	Class 330: Amplifiers
39 Other general electrical equipment	Class 318: Electricity, motive power systems
	Class 310: Electrical generator or motor structure
53 Other instruments and controls	Class 324: Electricity, measuring and testing

Table 7.15. The core Electrical / electronic technological sectors 1960-90

Technological Group	Patent Class / Subclass: Sector of Activity
34 Other electrical communication systems	Class 340: Communications, electrical
36 Image and sound equipment	Class 358: Pictorial communication; television
37 Illumination devices	Class 315: Electrical lamp and discharge devices, systems
	Class 313: Electrical lamp and discharge devices, consumable electrodes
38 Electrical devices and systems	Class 361: Electricity, electrical systems and devices
39 Other general electrical equipment	Class 437: Semiconductor device manufacturing: processes
40 Semiconductors	Class 357: Active solid state devices, eg transistors, solid state diodes
	Class 307: Electrical transmissions or interconnection systems
41 Office equipment and data processing systems	Class 364: Electrical computers and data processes systems
53 Other instruments and controls	Class 250: Radiant energy
	Class 324: Electrical, measuring and testing
	Class 350: Optics, systems and elements

Table 7.16. The core Mechanical technological sectors 1920-40

Technological Group	Patent Class / Subclass: Sector of Activity
6 Photographic chemistry	Class 430: Radiation imagery chemistry-processes, composition or products
14 Miscellaneous metal products	Class 220: Metallic receptacles
16 chemical and allied equipment	Class 51: Abrading
17 Metal working equipment	Class 72: Metal deforming
19 Building material processing equipment	Class 65: Glass manufacturing
21 Agricultural equipment	Class 56: Harvesters
	Class 172: Earth working
25 Textile and clothing machinery	Class 12: Boot and shoe making
	Class 112: Sewing
26 Printing and publishing machinery	Class 199: Type casting
50 Non metallic mineral products	Class 65: Glass manufacturing
52 Photographic equipment	Class 354: Photography

Table 7.17. The core Mechanical technological sectors 1960-90

Technological Group	Patent Class / Subclass: Sector of Activity
1 Food and tobacco products	Class 426: Food or edible material, processes, compositions and products
6 Photographic chemistry	Class 430: Radiation imagery chemistry-processes, composition or product
13 Metallurgical processes	Class 75: Metallurgy
14 Miscellaneous metal products	Class 220: Metallic receptacles
17 Metal working equipment	Class 72: Metal deforming
20 Assembly and material handling equipment	Class 414: Material or articles handling
28 Other specialised machinery	Class 425: Plastic articles or earthenware shaping or treating: apparatus
29 Other general industrial equipment	Class 137: Fluid handling
43 Motor vehicles	Class 180: Motor vehicles
50 Non-metallic mineral products	Class 501: Compositions: ceramic
	Class 65: Glass manufacturing
52 Photographic equipment	Class 354: Photography

Table 7.18. The core Transport technological sectors 1920-40

Technological Group	Patent Class / Subclass: Sector of Activity
29 Other general industrial equipment	Class 188: Brakes
	Class 192: Clutches and power stop control
42 Internal combustion engines	Class 123: Internal combustion engines
44 Aircraft	Class 244: Aeronautics
53 Other instruments and controls	Class 74: Machine elements and mechanisms

Table 7.19. The core Transport technological sectors 1960-90

Technological Group	Patent Class / Subclass: Sector of Activity
13 Metallurgical processes	Class 29: Metal working
29 Other general industrial equipment	Class 188: Brakes
	Class 192: Clutches and power stop control
	Class 137: Fluid handling
	Class 303: Fluid pressure brake and analogous systems
31 Power plants	Class 60: Power plants
42 Internal combustion engines	Class 123: Internal combustion engines
44 Aircraft	Class 244: Aeronautics
47 Other transport equipment	Class 280: Land vehicles
49 Rubber and plastic products	Class 152: Resilient tires and wheels
53 Other instruments and controls	Class 74: Machine elements and mechanisms
	Class 73: Measuring and testing

7.6.3. Trajectories of Corporate Technological Competencies

The firms in Tables 7.20-7.27 are sorted in accordance with their taxonomic position by the end of each technological epoch, i.e. 1940 and 1990. The tables allow for shifts in company names: i.e. different names may be subscribed to the same company-code at different points in time (the 1940 and 1990 names are listed here). However, although few companies also shifted sub-industrial grouping (e.g. Du Pont was always a mainstream chemicals firm before moving into petrochemicals more recently) only the recent (i.e. 1990) sub-industry is recorded in all cases).

Figure 7.7. Taxonomy of types of corporate technological competencies or profiles (a simplified version of Figure 7.3)

1 Niche peripheral followers	**2** Focused core followers	**3** Niche peripheral leaders	**4** Focused core leaders
5 Broad peripheral followers	**6** Broad core followers	**7** Broad peripheral leaders	**8** Broad core leaders

The TRAJECTORY column in Tables 7.20-7.27 illustrates the movement, if any, of the firm across the different taxonomic groupings. Please note that the positions of firms are merely in relation to the technological regime or paradigm in questions (see Tables 7.4-7.19 above), so e.g. a technological follower of a specific technological regime might well have very large competence within technologies not central to the regime (see section 7.2.2). All the firms selected for this analysis are major technological active corporations, both historically and at a global scale.

The distribution of firms' types of technological competencies: End of interwar period.

Table 7.20. Taxonomic distribution of Chemical firms in 1940

Taxonomic grouping / Industrial code and sector	Firm code	FIRM	TRAJECTORY 1930-1940
1: Niche peripheral followers /			
4: Chemicals n.e.s.	120	HERCULES (US)	5-1
4: Chemicals n.e.s.	168	RHONE-POULENC (FR)	1-1
13: Textiles and clothing	555	COURTAULDS (UK)	1-1
13: Textiles and clothing	918	BRITISH CELANESE (UK)	2-1
18: Coal and petroleum products	674	GULF OIL (US)	2-1
2: Focused core followers /			
4: Chemicals n.e.s.	166	SWISS-IG (SZ)	1-2
5: Pharmaceuticals	210	SCHERING (DE)	2-2
3: Niche peripheral leaders /			
4: Chemicals n.e.s.	112	UNION CARBIDE (US)	1-3
18: Coal and petroleum products	671	MOBIL (US)	1-3
5: Broad peripheral followers /			
4: Chemicals n.e.s.	111	DOW CHEMICAL (US)	6-5
4: Chemicals n.e.s.	117	AMERICAN CYANAMID (US)	2-5
4: Chemicals n.e.s.	279	DEGUSSA (DE)	5-5
18: Coal and petroleum products	673	ST. OIL OF CALIFORNIA (US)	1-5
6: Broad core followers /			
4: Chemicals n.e.s.	115	ALLIED (US)	6-6
4: Chemicals n.e.s.	139	ICI (UK)	2-6
4: Chemicals n.e.s.	805	IG FARBENINDUSTRIE (DE)	6-6
18: Coal and petroleum products	707	DU PONT (US)	8-6
7: Broad peripheral leaders /			
18: Coal and petroleum products	670	STANDARD OIL NEW JERSEY (US)	4-7
18: Coal and petroleum products	672	TEXACO (US)	7-7
18: Coal and petroleum products	675	STANDARD OIL:INDIANA (US)	1-7

Table 7.21. Taxonomic distribution of Electrical/electronic firms in 1940

Taxonomic grouping / Industrial code and sector	Firm code	FIRM	TRAJECTORY 1930-1940
1: Niche peripheral followers /			
8: Electrical equipment n.e.s.	392	SINGER (US)	1-1
8: Electrical equipment n.e.s.	407	GENERAL ELECTRIC CO (UK)	1-1
9: Office equipment	442	IBM (US)	1-1
9: Office equipment	445	HONEYWELL (US)	1-1
2: Focused core followers /			
8: Electrical equipment n.e.s.	354	NORTHERN ENGINEERING INDUS (UK)	1-2
3: Niche peripheral leaders /			
8: Electrical equipment n.e.s.	384	ITT (US)	4-3
4: Focused core leaders /			
8: Electrical equipment n.e.s.	428	AEG-TELEFUNKEN (DE)	4-4
8: Electrical equipment n.e.s.	438	BROWN BOVERI (SZ)	3-4
8: Electrical equipment n.e.s.	873	RADIO CORPORATION OF AMERICA (US)	4-4
6: Broad core followers /			
8: Electrical equipment n.e.s.	383	GENERAL ELECTRIC (US)	6-6
8: Electrical equipment n.e.s.	386	WESTINGHOUSE ELECTRIC (US)	5-6
8: Electrical equipment n.e.s.	872	AMERICAN TELEPHONE AND TELEGRAPH CO (US)	6-6
7: Broad peripheral leaders /			
8: Electrical equipment n.e.s.	427	SIEMENS (DE)	7-7

Table 7.22. Taxonomic distribution of Mechanical firms in 1940

Taxonomic grouping / Industrial code and sector	Firm code	FIRM	TRAJECTORY 1930-1940
1: Niche peripheral followers /			
7: Mechanical engineering n.e.s.	366	GUTEHOFFNUNGSHUTTE (DE)	3-1
7: Mechanical engineering n.e.s.	372	SKF (SE)	2-1
7: Mechanical engineering n.e.s.	382	SCHNEIDER (FR)	1-1
7: Mechanical engineering n.e.s.	846	WESTINGHOUSE AIR BRAKE	1-1
7: Mechanical engineering n.e.s.	857	RHEINMETALL / RHEINMETALL BORSIG (DE)	3-1
2: Focused core followers /			
7: Mechanical engineering n.e.s.	326	DEERE (US)	4-2
17: Non-metallic mineral products	646	NORTON (US)	8-2
20: Other manufacturing	775	OWENS ILLINOIS (US)	2-2
3: Niche peripheral leaders			
7: Mechanical engineering n.e.s.	357	VICKERS-ARMSTRONG (UK)	1-3
19: Professional and scientific instruments	771	CARL-ZEISS-STIFTUNG (DE)	3-3
4: Focused core leaders /			
20: Other manufacturing	777	CORNING GLASS WORKS (US)	4-4
5: Broad peripheral followers /			
6: Metals	215	BETHLEHEM STEEL (US)	5-5
6: Metals	275	KRUPP (DE)	5-5
7: Mechanical engineering n.e.s.	336	ALLIS-CHALMERS (US)	5-5
7: Mechanical engineering n.e.s.	853	BAKER PERKINS CO INC (UK)	5-5
6: Broad core followers /			
6: Metals	213	US STEEL (US)	6-6
6: Metals	217	CONTINENTAL GROUP (US)	2-6
6: Metals	220	AMERICAN CAN (US)	6-6
7: Mechanical engineering n.e.s.	325	INTERNATIONAL HARVESTOR (US)	2-6
7: Mechanical engineering n.e.s.	335	EMHART (US)	6-6
7: Broad peripheral leaders /			
1: Food	9	BORDEN (US)	5-7
7: Mechanical engineering n.e.s.	373	ALFA-LAVAL (SE)	3-7
8: Broad core leaders /			
19: Professional and scientific instruments	760	EASTMAN KODAK (US)	4-8

Table 7.23. Taxonomic distribution of Transport firms in 1940

Taxonomic grouping / Industrial code and sector	Firm code	FIRM	TRAJECTORY 1930-1940
1: Niche peripheral followers /			
11: Aircraft	515	UNITED TECHNOLOGIES (US)	1-1
2: Focused core followers /			
10: Motor vehicles	494	DAIMLER-BENZ (DE)	6-2
10: Motor vehicles	498	ZAHNRADFABRIK FRIEDRICHSHAFEN (DE)	2-2
11: Aircraft	517	ROCKWELL INTERNATIONAL (US)	2-2
3: Niche peripheral leaders /			
10: Motor vehicles	496	BOSCH (DE)	1-3
16: Rubber and plastic products	633	DUNLOP (UK)	7-3
5: Broad peripheral followers			
16: Rubber and plastic products	629	FIRESTONE TIRE AND RUBBER (US)	5-5
6: Broad core followers /			
10: Motor vehicles	465	BENDIX (US)	2-6
7: Broad peripheral leaders /			
10: Motor vehicles	462	GENERAL MOTORS (US)	6-7
10: Motor vehicles	464	CHRYSLER (US)	6-7
16: Rubber and plastic products	628	GOODYEAR TIRE AND RUBBER (US)	5-7
16: Rubber and plastic products	630	BF GOODRICH (US)	5-7
8: Broad core leaders /			
10: Motor vehicles	463	FORD MOTOR (US)	8-8

The distribution of firms' types of technological competencies: End of recent period

Table 7.24. Taxonomic distribution of Chemical firms in 1990

Taxonomic grouping / Industrial code and sector	Firm code	FIRM	TRAJECTORY 1960-1970-1980-1990
1: Niche peripheral followers /			
4: Chemicals n.e.s.	64	UNILEVER (UK/NL)	1-3-1-1
4: Chemicals n.e.s.	116	W R GRACE (US)	8-6-2-1
4: Chemicals n.e.s.	120	HERCULES (US)	1-2-1-1
5: Pharmaceuticals	177	PROCTER & GAMBLE (US)	1-1-1-1
5: Pharmaceuticals	184	MERCK (US)	2-4-2-1
13: Textiles and clothing	555	COURTAULDS (UK)	1-1-1-1
13: Textiles and clothing	568	SNIA VISCOSA (IT)	1-1-1-1
18: Coal and petroleum products	672	TEXACO (US)	7-3-3-1
18: Coal and petroleum products	676	ATLANTIC RICHFIELD (US)	5-3-2-1
2: Focused core followers /			
4: Chemicals n.e.s.	139	ICI (UK)	2-5-5-2
4: Chemicals n.e.s.	159	BAYER (DE)	3-3-1-2
4: Chemicals n.e.s.	166	CIBA-GEIGY (SZ)	2-2-2-2
4: Chemicals n.e.s.	279	DEGUSSA (DE)	2-2-2-2
5: Pharmaceuticals	204	SANDOZ (SZ)	3-1-2-2
5: Pharmaceuticals	210	SCHERING (DE)	1-2-2-2
18: Coal and petroleum products	671	MOBIL (US)	2-6-2-2
3: Niche peripheral leaders /			
4: Chemicals n.e.s.	115	ALLIED (US)	6-4-7-3
4: Chemicals n.e.s.	140	BOC GROUP (UK)	3-3-3-3
4: Chemicals n.e.s.	163	AGA (SE)	3-3-3-3
4: Chemicals n.e.s.	170	L'AIR LIQUIDE (FR)	1-1-3-3
5: Pharmaceuticals	203	ROCHE/SAPAC (SZ)	4-4-3-3
5: Broad peripheral followers /			
4: Chemicals n.e.s.	112	UNION CARBIDE (US)	3-7-3-5
4: Chemicals n.e.s.	158	HOECHST (DE)	5-5-5-5
18: Coal and petroleum products	707	DU PONT (US)	6-5-5-5
18: Coal and petroleum products	709	BP (UK)	5-5-5-5

Table 7.24 Continued.

6: Broad core followers /			
4: Chemicals n.e.s.	111	DOW CHEMICAL (US)	6-6-6-6
4: Chemicals n.e.s.	114	MONSANTO (US)	2-6-8-6
4: Chemicals n.e.s.	117	AMERICAN CYANAMID (US)	6-6-6-6
4: Chemicals n.e.s.	119	CELANESE (US)	5-1-1-6
4: Chemicals n.e.s.	160	BASF (DE)	6-6-6-6
4: Chemicals n.e.s.	168	RHONE-POULENC (FR)	2-2-6-6
18: Coal and petroleum products	673	ST. OIL OF CALIFORNIA (US)	7-6-6-6
18: Coal and petroleum products	674	GULF OIL (US)	7-6-6-6
18: Coal and petroleum products	678	PHILLIPS PETROLEUM (US)	5-5-5-6
18: Coal and petroleum products	708	SHELL (UK/NL)	6-6-6-6
7: Broad peripheral leaders /			
4: Chemicals n.e.s.	113	3M'S (US)	7-7-3-7
8: Broad core leaders /			
18: Coal and petroleum products	670	EXXON (US)	6-6-6-8
18: Coal and petroleum products	675	STANDARD OIL:INDIANA (US)	5-6-6-8

Table 7.25. Taxonomic distribution of Electrical/electronic firms in 1990

Taxonomic grouping / Industrial code and sector	Firm code	FIRM	TRAJECTORY 1960-1970-1980-1990
1: Niche peripheral followers /			
8: Electrical equipment n.e.s.	354	NORTHERN ENGINEERING INDUS (UK)	4-2-1-1
8: Electrical equipment n.e.s.	392	SINGER (US)	1-1-1-1
8: Electrical equipment n.e.s.	409	BICC (UK)	3-1-1-1
8: Electrical equipment n.e.s.	428	AEG-TELEFUNKEN (DE)	1-2-1-1
8: Electrical equipment n.e.s.	437	LM ERICSSON (SE)	1-1-1-1
8: Electrical equipment n.e.s.	438	BROWN BOVERI (SZ)	3-1-1-1
2: Focused core followers /			
8: Electrical equipment n.e.s.	873	RADIO CORPORATION OF AMERICA (US)	6-6-6-2
5: Broad peripheral followers /			
8: Electrical equipment n.e.s.	386	WESTINGHOUSE ELECTRIC (US)	5-5-5-5
8: Electrical equipment n.e.s.	872	AMERICAN TELEPHONE AND TELEGRAPH CO (US)	8-6-6-5
6: Broad core followers /			
8: Electrical equipment n.e.s.	408	THORN EMI (UK)	2-2-2-6
9: Office equipment	445	HONEYWELL (US)	1-1-2-6
7: Broad peripheral leaders /			
8: Electrical equipment n.e.s.	383	GENERAL ELECTRIC (US)	5-7-5-7
8: Electrical equipment n.e.s.	384	ITT (US)	6-6-5-7
8: Electrical equipment n.e.s.	431	GENERALE D'ELECTRICITE (FR)	1-3-7-7
8: Broad core leaders /			
8: Electrical equipment n.e.s.	407	GENERAL ELECTRIC CO (UK)	8-8-6-8
8: Electrical equipment n.e.s.	427	SIEMENS (DE)	8-8-8-8
8: Electrical equipment n.e.s.	434	PHILIPS (NL)	8-8-8-8
9: Office equipment	442	IBM (US)	4-4-8-8

Table 7.26. Taxonomic distribution of Mechanical firms in 1990

Taxonomic grouping / Industrial code and sector	Firm code	FIRM	TRAJECTORY 1960-1970-1980-1990
1: Niche peripheral followers /			
6: Metals	213	US STEEL (US)	6-6-5-1
7: Mechanical engineering n.e.s.	335	EMHART (US)	1-1-1-1
7: Mechanical engineering n.e.s.	368	KHD (DE)	4-1-1-1
7: Mechanical engineering n.e.s.	372	SKF (SE)	1-1-1-1
7: Mechanical engineering n.e.s.	373	ALFA-LAVAL (SE)	1-1-1-1
7: Mechanical engineering n.e.s.	374	ATLAS COPCO (SE)	1-1-1-1
7: Mechanical engineering n.e.s.	846	AMERICAN STANDARD	3-3-3-1
14: Paper products n.e.s.	574	INTERNATIONAL PAPER (US)	1-1-1-1
17: Non-metallic mineral products	646	NORTON (US)	1-5-1-1
2: Focused core followers /			
1: Food	3	GENERAL FOODS (US)	2-2-2-2
1: Food	6	GENERAL MILLS (US)	5-5-6-2
6: Metals	217	CONTINENTAL GROUP (US)	6-6-6-2
7: Mechanical engineering n.e.s.	326	DEERE (US)	1-1-1-2
7: Mechanical engineering n.e.s.	366	GUTEHOFFNUNGSHUTTE (DE)	2-2-2-2
7: Mechanical engineering n.e.s.	853	BAKER PERKINS CO INC (UK)	6-2-2-2
3: Niche peripheral leaders /			
6: Metals	843	BOFORS GULLSPANG AB (SE)	1-3-3-3
7: Mechanical engineering n.e.s.	325	INTERNATIONAL HARVESTOR (US)	5-5-1-3
7: Mechanical engineering n.e.s.	376	SULZER BROS (SZ)	1-1-3-3
4: Focused core leaders /			
17: Non-metallic mineral products	653	PILKINGTON BROS (UK)	2-4-4-4
19: Professional and scientific instruments	771	CARL-ZEISS-STIFTUNG (DE)	4-4-4-4
19: Professional and scientific instruments	951	ERNST LEITZ OPTISCHE WERKE (DE)	4-4-4-4
5: Broad peripheral followers /			
1: Food	1	DART AND KRAFT (US)	2-7-5-5
1: Food	9	BORDEN (US)	4-3-5-5
6: Metals	277	METALLGESELLSCHAFT (DE)	2-2-1-5
6: Metals	288	PECHINEY UGINE KUHLMANN (FR)	2-3-3-5
7: Mechanical engineering n.e.s.	336	ALLIS-CHALMERS (US)	5-5-5-5
7: Mechanical engineering n.e.s.	371	ASEA (SE)	1-1-1-5

Table 7.26 Continued

6: Broad core followers /			
6: Metals	215	BETHLEHEM STEEL (US)	6-6-6-6
6: Metals	220	AMERICAN CAN (US)	6-6-6-6
6: Metals	222	REYNOLDS METALS (US)	6-6-6-6
6: Metals	276	MANNESMANN (DE)	7-7-8-6
7: Mechanical engineering n.e.s.	324	CATERPILLAR TRACTORS (US)	2-2-6-6
17: Non-metallic mineral products	658	SAINT-GOBAIN (FR)	8-8-6-6
7: Broad peripheral leaders /			
7: Mechanical engineering n.e.s.	329	FMC (US)	7-7-7-7
7: Mechanical engineering n.e.s.	352	HAWKER SIDDELEY (UK)	3-3-7-7
7: Mechanical engineering n.e.s.	357	VICKERS (UK)	5-8-3-7
14: Paper products n.e.s.	577	KIMBERLEY CLARK (US)	7-7-7-7
8: Broad core leaders /			
19: Professional and scientific instruments	760	EASTMAN KODAK (US)	6-8-8-8
20: Other manufacturing	775	OWENS ILLINOIS (US)	6-6-8-8
20: Other manufacturing	777	CORNING GLASS WORKS (US)	2-8-8-8

Table 7.27. Taxonomic distribution of Transport firms in 1990

Taxonomic grouping / Industrial code and sector	Firm code	FIRM	TRAJECTORY 1960-1970-1980-1990
1: Niche peripheral followers /			
11: Aircraft	515	UNITED TECHNOLOGIES (US)	1-3-3-1
11: Aircraft	531	ROLLS ROYCE (UK)	1-2-2-1
11: Aircraft	914	CURTISS AIRPLANE CO / CURTISS WRIGHT / WRIGHT AERONAUTICAL CORP (US)	2-2-1-1
16: Rubber and plastic products	628	GOODYEAR TIRE AND RUBBER (US)	6-8-1-1
10: Motor vehicles	496	BOSCH (DE)	1-2-2-1
2: Focused core followers /			
10: Motor vehicles	494	DAIMLER-BENZ (DE)	2-2-2-2
10: Motor vehicles	502	PEUGEOT (FR)	2-2-2-2
10: Motor vehicles	509	ROVER (UK)	2-2-2-2
16: Rubber and plastic products	633	DUNLOP (UK)	6-6-2-2
3: Niche peripheral leaders /			
11: Aircraft	518	McDONNELL DOUGLAS (US)	5-7-7-3
16: Rubber and plastic products	629	FIRESTONE TIRE AND RUBBER (US)	2-2-3-3
16: Rubber and plastic products	630	BF GOODRICH (US)	3-3-3-3
16: Rubber and plastic products	632	UNIROYAL (US)	3-3-3-3
4: Focused core leaders /			
11: Aircraft	516	BOEING (US)	2-8-4-4
11: Aircraft	530	BRITISH AEROSPACE (UK)	2-2-4-4
6: Broad core followers /			
10: Motor vehicles	462	GENERAL MOTORS (US)	5-5-5-6
10: Motor vehicles	464	CHRYSLER (US)	6-6-6-6
10: Motor vehicles	465	BENDIX (US)	6-6-8-6
10: Motor vehicles	466	BORG-WARNER (US)	6-2-5-6
10: Motor vehicles	501	RENAULT (FR)	4-6-6-6
10: Motor vehicles	511	LUCAS (UK)	6-6-6-6
7: Broad peripheral leaders /			
10: Motor vehicles	463	FORD MOTOR (US)	7-7-7-7

8. Trajectories of Corporate Technological Leadership: Implications for Innovation Diffusion in the Course of Growth[1]

Abstract

This chapter examines the dispersion of corporate technological leadership in technological sectors with rapid development historically. It is shown that the technological diffusion (or catching up) processes of innovation is positively related to the rate of development. However, during such phases of fast growth, 'leaders' tend to perceive at least some elements of their leading position. It is only over longer historical periods, when technological opportunities change (and thus the underlying competitive landscape alters) that the actual composition of corporate technological leadership tends to shift significantly. The results have implications for how we need to understand technological diffusion or expansion processes inseparably from development processes or rate of advance, and *vice versa*.

8.1. INTRODUCTION: IN THE SCHUMPETERIAN VEIN

The analysis in this chapter will concentrate in more detail on the evolution of corporate technological leadership which is crucial in Schumpeter's theory. As Schumpeter's specific contribution to business cycle theory (as argued by Kuznets 1940) deals with the association between the 'distribution of entrepreneurial ability or leadership' and 'cyclical fluctuations in the rate of innovation (Chapter 4)', this chapter analyses the extent to which technological leadership is concentrated in relatively few firms, or is more widely dispersed, in different epochs of technological development in the interwar period and in the period since 1960. It also examines whether the

[1] This chapter is a revised version of Cantwell and Andersen (1996).

composition of leading firms is relatively stable or prone to fluctuate over time, with reference to corporate presence in selected fields of technological activity in the forefront of new development in each period. To keep in accord with a Schumpeterian story, particular attention is given to investigating innovation diffusion of corporate technological leadership in the course of technological growth. Thus, focus is here put on those firms which in Chapter 7's taxonometric grouping have been termed technological leaders.

Corporate technological leadership is defined as in Chapter 7. That is, among the largest firms in each industry the 'leaders' are those that are particularly specialised in the fastest growing areas of new technology (i.e. areas of greatest opportunities). In its turn, corporate technological specialisation is an indicator of each firm's distinct fields of technological competence. Hence, as identified in Chapter 7, the leading firms in each major international industry in any period are those firms which have established a competence in the areas of the most rapid technological development within a given technological epoch or paradigm, measured by the rate of growth of patenting.

As in Chapter 7, this chapter is restricted to considering intra-industry technological diffusion - i.e. the erosion of technological leadership within a broadly defined industry - and does not consider inter-industry diffusion. The justification for analysing the structure of corporate patenting as an intra-industry issue was also presented in Chapter 7 (section 7.1). Furthermore, Chapter 6 also indicated that industries still tend to patent within their own fields (i.e. Chemical firms tend to patent within chemical technologies, and so forth), although evidence also showed how blurring of industry boundaries has become more apparent over time.

The overall analysis is considered in relation to two technological epochs, 1920-40 (the interwar period) and 1960-90 (the recent period), as identified in Chapter 3 and also applied in Chapter 7. The period in the middle will be cut out of the analysis (i) due to the effects of the Second World War which created fluctuations in patenting activity, and (ii) due to a technological paradigm shift (identified in Chapter 3).

8.1.1. Chapter Outline

First the data used for this chapter will be selected, and then the theoretical framework concerning technological leadership issues will follow. This will be followed by an identification of the actual composition of companies that played a leading role in the development of some particular fast growing technologies across different time periods.

The next and major issue addressed in this chapter is an analysis of the way in which leadership disperses as technologies advance, which is believed to be linked to innovation diffusion or 'catching up'. Particular attention is paid to what might be termed a Schumpeterian hypothesis; namely, that technological activity tends to be highly concentrated in relatively few firms when a new area of technological activity appears, but then a dispersion of leadership occurs as the growth of patenting in that area takes off. This is an application of the logistic idea of technological development (identified in Chapter 4), and change in the intensity of leadership over such a technological cycle.

Therefore, it seems reasonable to expect that the maximum rate of growth of patenting in a technological field is associated with the fastest dilution of corporate technological leadership. The question then becomes the extent to which technological leadership dissipates in the course of technological development. This might well vary across technological fields as well as depend on the industry in question. However, due to the historical shift towards more integrated technological systems through the fusion of diverse and formerly separate branches of technology (as identified in Chapter 3 and Andersen 1998), an overall historical trend of decline in the concentration of technological leadership in any individual sector is to be expected. Hence, the degree of concentration of activity across the largest firms in particular fast growing fields over time is analysed, with reference to the relationship between changes in such concentration and the overall growth of patenting in the field in question.

Finally, the extent to which existing leaders incrementally consolidate their lead as new technological fields emerge, or at least preserve rather than lose their position, will be tested. Both the average distance between the leaders and followers in each industry in the fastest growing fields over time (i.e. the degree of catching up), and the extent of shifts in the actual composition of corporate technological leadership will be tested.

8.1.2. Data Selection

On similar lines as Chapter 7, this Chapter exploits some of the panel data characteristics that patent statistics can provide when classified by (i) type of technological activity at the very detailed 399 disaggregated class level, (ii) corporate ownership when classified at the 284 corporate group level, (iii) broad industry of firm, as well as (iv) over time. The organisation of the 399 technological sectors and 284 corporate groups are presented in Chapter 2. The analysis allows as well for corporate groups that operated under different names during different periods. The technological and industrial sectors are sorted into their corresponding Chemical, Electrical/electronics, Mechanical and Transport broad groups (as identified in Chapter 2). Hence, Non-industrial technologies are left out, as there is no counterpart industrial group. In the design of empirical categories and methods, further restrictions have been imposed when selecting technological sectors and corporate patenting for specific analysis. They are presented, where appropriately, in the text.

However, one specific issue will be addressed here. As noted earlier in Chapter 7, corporate technological leadership is defined as a specialisation by a firm relative to others in its industry in the fastest growing sectors of activity that are most relevant to technological development in that industry. In Chapter 7 the areas of greatest technological opportunities were selected by dividing fields into the four broad technological groups (Chemical, Electrical/electronic, Mechanical and Transport). Then each group was ranked by the rate of growth of their patent stock, and the areas of greatest opportunities were then selected. This was done by taking the 12 fastest growing classes within each broad technological group in each technological epoch - 1920-40 and 1960-90 - which is considered for this analysis, as described in section 8.1. However, only 6 fast growing patent classes were selected for transport technologies to avoid problems with small sectors, owing to the few sectors within that technological group. Hence, the 42 areas (in total for each period) of greatest opportunities selected in Chapter 7 are also used for the purpose of this chapter, and they were set out in Tables 7.4-7.11 (Appendix section 7.6.1).

8.2. THEORETICAL FRAMEWORK: TECHNOLOGICAL LEADERSHIP AND THE IMPLICATION OF SIZE

Using the accumulated patent stock for each firm and for the total of all firms in the relevant broad industrial group, each firm's pattern of technological specialisation is measured by an index termed revealed technological advantage (RTA). As mentioned earlier (Chapter 7), the index is calculated for each firm relative to others in the same broad industrial group across the various sectors of technological activity. Hence, the RTA of a firm in a particular fast growing sector (i.e. sector with high opportunity, see data selection - section 8.1.2 - above) is given by the firm's share in that sector of US patents granted to companies in the same broad industrial group, relative to its overall share of all US patents granted to companies in the same broad industry. Denoting by P_{ij} for firm i in a particular industry, the number of US patents granted in sector j, the RTA index for each firm in that industry is defined as follows:

(i)

$$RTA_{ij} = \frac{P_{ij} / \sum_i P_{ij}}{\sum_j P_{ij} / \sum_{ij} P_{ij}}$$

The index varies around unity, such that values greater than unity suggest that a firm is positively or comparatively specialised in technological development in the sector in question.

For this study the RTA index has been calculated for the firms of each broad industry for each year from 1920 to 1990.

However, it ought to be said that the RTA index is open to an interpretation that would provide an exaggerated or somewhat distorted measurement of corporate technological leadership, in our sense of a specialisation in the fastest growing fields of activity. To understand why, we must distinguish firms by their (relative) size, as well as by whether they are technological leaders or followers. While all the firms in this study are large (or quite quickly became large), some are larger than others, and these can be thought of as giant corporations (such as the US firm General Electric or the German company IG Farben) by comparison with other smaller firms in the investigation. As mentioned in Chapter 7, it is well known that diversification is a positive function of firm size, so that smaller firms are generally more highly specialised, and giant corporations less so (Chandler 1990). Such evidence has also more recently been confirmed empirically by Cantwell and colleagues (Cantwell and Bachmann, 1998, Cantwell and Fai 1999, Cantwell and Santangelo 2000).

Hence what matters in this respect is neither their overall size nor their overall degree of specialisation, but rather the actual composition or profile of that specialisation - that is, whether their resources are especially geared towards development in the fastest growing fields. As corporate technological leadership is defined here, smaller firms may be leaders if they are particularly oriented towards development in fastest growing areas, although the absolute level of their activity in these fields may be less than that of giant companies which specialise instead in mature technologies and so act as followers.

However, a problem arises when examining one particular technological sector of activity in isolation; when, for example, comparing a giant corporation and a smaller firm that share a specialisation in that field, the RTA value of the smaller firm tends to be larger because its overall degree of specialisation is narrower. As this study examines each of the individual fast growing sectors separately in turn, and because of the possible influence of the size of the firms on the cross-company distribution of the RTA index in this context, it is as well to exclusively take into account also the size (associated with the cross-firm distribution) of the absolute level of technological activity. This is achieved through considering the patent share

(PS) of each firm, which is the numerator of the RTA index, and thus reflects the absolute level of activity of the firm in a sector, without normalising for its overall size. The PS of a firm in a particular technological sector is the firm's share in that sector of US patents granted to companies in the same broad industry.

Where P_{ij} denotes the US patents granted in sector j to firm i within a particular industry, the PS index for each firm in each industry is defined as follows:

(ii) $$PS_{ij} = \frac{P_{ij}}{\sum_i P_{ij}}$$

The PS index has been calculated for the firms of each broad industry for each year from 1920 to 1990.

As the PS index is more a reflection of size than of technological leadership, there is no reason to expect any particular link between sectoral growth and declining concentration within each industry. However, as mentioned in the introduction to this Chapter, a longer term historical decline in cross-firm concentration in both the PS and the RTA index is expected, as firms have become more technologically diversified (Chapter 6) due to their own growth, and due to an increasing technological interrelatedness between previously autonomous fields of activity (Chapter 3).

It might be argued that in considering the fastest growing fields, leadership should be defined as entailing a high absolute level of activity rather than a relative specialisation in these fields; that is, leadership should be defined with reference to the PS index and not the RTA index. However, it is preferred here to restrict the terminology of leadership to the relative specialisation of firms in fast growing fields, since this book seeks to attempt to capture leading contributions to technological change (in the sense of the transformation of products and processes) and not the leading contributions to the level of research or knowledge-generating activity for its own sake. That is, when considering the capacity for innovation in firms, the level of technological activity must be taken relative to the size of the production base to which research relates. Giant corporations have a higher absolute level of both research (in most or all technological fields) and production, but they have only a weak technological competence (a weak ability to innovate) in fields in which they are not specialised. Hence, the RTA index, rather then the PS index, measures which firms have a particular technological competence in a field, as opposed to a high level of research activity.

This view of leadership in terms of competence and hence the contribution of firms to innovation can be supported with reference to the literature on the history of business technology. For example, by the late 1930s the large chemical firms Du Pont and ICI both carried out a substantial volume of research activity in the area of dyestuffs (organic chemicals). However, both Du Pont and ICI had low RTA values in dyestuffs despite a significant PS, and they both remained followers and not leaders in this field, in the sense that they were not generally capable of initiating development in an area in which they were not particularly competent relative to their major rival, IG Farben (Reader 1975; Hughes 1989).

8.3. CORPORATE TECHNOLOGICAL TOP-LEADERS AND LARGEST CONTRIBUTORS

As described earlier, the rate of growth of patenting across different fields in each period serves as an indicator of the areas in which technological opportunities were greatest at that time. The fields of the greatest technological opportunities change from one period to the next. However, as illustrated in Tables 7.4-7.11 (Chapter 7, Appendix, section 7.6.1), there are a few sectors (10 in total) that have continued to be sources of important opportunities and have witnessed a sustained increase in patenting over the entire period from 1920-40 to 1960-90. The firms that were particularly specialised in four such sectors - one from each of the major categories (i.e. Chemicals, Electrical/electronics, Mechanical, Transport) of activity - are identified in Table 8.1, while the firms that made the greatest absolute contribution of development in the same areas are shown in Table 8.2. The value of the index of specialisation (RTA) or the size index (PS) is recorded for the leading firms in each sector, in 1940 and in 1990. Only companies with an accumulated patent stock of at least 200 patents at the end of each broad period, that is in 1940 and 1990 respectively, are included in these tables, since firms with a lower level of patenting tend to obtain high RTA values in a sector on the basis of small numbers of patents, which may unduly reflect chance factors. Furthermore, firms that are technological leaders in *both* the interwar and recent period are shown in 'italics', and so are firms that make a large contribution in *both* the interwar and recent period.

Table 8.1 illustrates that over very long periods (50 years) positions of technological leadership in specific fields tend to be at least somewhat eroded or transformed. However, even over this length of time there are occasionally firms or corporate groups that retain their leading roles. This applies to Philips within 'active solid state devices'; Carl-Zeiss-Stiftung and Corning Glass Works within 'optics'; and Bosch and Chrysler in 'internal combustion engines'.

Table 8.1. The top-five technological leaders for selected technological sectors over time

	interwar (1940)		recent (1990)	
Technological activity	Firm	RTA	Firm	RTA
CHEMICAL	Monsanto	3.64	Boehringer Ingelheim	7.12
Group: Pharmaceuticals and biotechnology	Dow Chemical	2.74	Beecham Group	6.74
Sector 514: Drug, bio-affecting and body treating compositions	American Cyanamid	2.13	Wellcome Foundation	6.67
	Gulf Oil	1.71	Merck	4.97
	Standard Oil: Indiana	1.53	Glaxo	4.85
ELECTRICAL	*Philips*	16.48	AEG-Telefunken	2.34
Group: Semiconductors	ITT	4.96	Brown Boveri	1.84
Sector 357: Active solid state devices, e.g.	Westinghouse Electric	1.31	RCA	1.79
Transistors, solid state diodes	General Electric	1.19	IBM	1.61
			Philips	1.50
MECHANICAL	*Carl-Zeiss-Stiftung*	27.93	Ernst Leitz Optische Werke	20.43
Group: Other instruments and controls	*Corning Glass Works*	1.96	*Carl-Zeiss-Stiftung*	20.20
Sector 350: Optics, systems	Eastman Kodak	1.89	Pilkington Bros	10.50
and elements			*Corning Glass Works*	3.94
			W C Heraeus	2.47
TRANSPORT	*Bosch*	3.56	*Bosch*	4.24
Group: Internal combustion engines	United Aircraft	1.80	Renault	2.26
Sector 123: Internal	General Motors	1.19	Daimler-Benz	2.14
Combustion engines	*Chrysler*	1.07	Lucas	1.93
			Chrysler	1.74

Source: Cantwell and Andersen 1996

Table 8.2. The largest contributors to development in selected technological sectors over time

	interwar (1940)		recent (1990)	
Technological activity	Firm	PS %	Firm	PS %
CHEMICAL	*IG Farben-Industrie*	23.92	*Bayer*	11.50
Group: Pharmaceuticals and biotechnology	Du Pont	16.44	Merck	10.55
Sector 514: Drug, bio-affecting and body treating compositions	Dow Chemical	11.86	Ciba-Geigy	8.86
	Standard Oil: Indiana	5.19	*Hoechst*	8.68
	Monsanto	5.09	Smith Kline Beckman	4.81
ELECTRICAL	*Philips*	16.34	IBM	15.90
Group: Semiconductors	AT&T	9.49	*Philips*	15.20
Sector 357: Active solid state	ITT	7.43	General Electric	12.05
Devices, e.g. Transistors, solid state diodes	*RCA*	6.36	*RCA*	11.48
	Siemens	1.25	*Siemens*	11.18
MECHANICAL	*Carl-Zeiss-Stiftung*	59.09	*Eastman Kodak*	24.27
Group: Other instruments and controls	*Eastman Kodak*	21.36	*Carl-Zeiss-Stiftung*	21.78
Sector 350: Optics, systems and elements	*Corning Glass Works*	2.59	*Corning Glass Works*	20.73
			Ernst Leitz Optische Werke	10.53
			Pilkington Bros	7.42
TRANSPORT	*General Motors*	31.80	*Bosch*	37.19
Group: Internal combustion engines	Bendix	24.60	*General Motors*	18.36
Sector 123: Internal Combustion engines	*Bosch*	7.93	Ford Motor	8.59
	United Aircraft	6.15	Lucas	8.22
	Chrysler	5.45	Daimler-Benz	6.68

Source: Cantwell and Andersen 1996

It is still more common for firms to sustain a top five ranking in terms of their absolute contribution to activity in a sector, as shown in Table 8.2. This applies to IG Farben-Industrie and its successors within 'pharmaceuticals'; Philips and Siemens within 'active solid state devices'; Carl-Zeiss-Stiftung, Eastman Kodak and Corning Glass Works in 'optics'; and General Motors and Bosch in 'engines'. Again, this impression of some stability in leadership is confirmed by regressions of the cross-firm PS index in 1990 on 1940, which show a significant correlation at the 5 per cent level of correlation in all of the ten fields that were fast growing in both historical periods.

8.4. TESTING THE SCHUMPETERIAN HYPOTHESIS OF INNOVATION DIFFUSION IN THE COURSE OF TECHNOLOGICAL GROWTH.

For each of the fast growing classes of activity selected, the degree of cross-firm concentration of corporate technological leadership in any year is measured by the coefficient of variation (CV) of the index of corporate technological specialisation (RTA) across all the firms in the industry in question: $[CV_{RTA} = (\sigma_{RTA}/\mu_{RTA})\cdot 100\%]$.[2] Higher values of this measure denote that technological leadership is concentrated in relatively fewer firms, whereas lower values of this measure denote that technological leadership is less concentrated. For the purposes of comparison this chapter also considers the degree of concentration in the size distribution of activity in the same fast growing sectors, as measured by the coefficient of variation of the PS index $[CV_{PS} = (\sigma_{PS}/\mu_{PS})\cdot 100\%]$. Of course, in this context the term 'size' refers to the technological size of the firm as measured by its volume of patenting, and not size as conventionally measured by the value of sales or the volume of employment; although all these size measures are correlated across firms (Cantwell and Sanna-Randaccio 1993).

Table 8.3 shows how the degree of concentration of technological leadership has tended to fall historically. When considering the interwar period we see that of the 41 sectors in which the largest US and European firms had become involved by 1920, there was a general trend towards a decline in concentration in 27 such sectors, while there were 13 sectors in which the measure of corporate concentration rose and one in which it

[2] Of course, other measures of concentration might have been used, such as the concentration ratio or the Herfindahl index. However, as mentioned in Chapter 3, for a given number of firms there is a strict relationship between the coefficient of variation and the Herfindahl index.

remained stable. Of the 13 sectors in which the measure of corporate concentration rose, six were in the Mechanical group, four were chemical sectors, two were in Transport, and one in electrical equipment. However, the measure of concentration rose only very slightly (proportionately) in all these sectors, except in four areas of mechanical technology - namely in sector 148 in 'Metallurgical processes', and sectors 350, 352 and 356 in 'Other instruments and controls'. These appear to be sectors in which cumulative learning, based on an early start, seems to have paid particular dividends in the interwar years. In the recent period we see that, of the 40 sectors in which the largest US and European firms had become involved by 1960, all except one have experienced decreasing concentration of technological leadership. The one sector with slightly increasing concentration is 114 (ships and marine propulsion) in the Transport group.

Table 8.3. The change in cross-firm concentration of corporate technological leadership and in the size distribution of corporate technological activity over time (by numbers of sectors)

	interwar (1920-40)		recent (1960-90)	
Technological group	Concentration falls	Concentration rises or is stable	Concentration falls	Concentration rises or is stable
Change in Concentration of Corporate Technological Leadership (RTA)				
Chemical	7	5	12	0
Electrical*	10	1	12	0
Mechanical**	6	6	10	0
Transport	4	2	5	1
Change in Concentration of the Size Distribution of Corporate Technological Activity (PS)				
Chemical	11	1	12	0
Electrical*	10	1	11	1
Mechanical**	11	1	9	1
Transport	4	2	5	1

**: No corporate patenting in 1920 in one technological sector out of those selected for the interwar period.*

***: No corporate patenting in 1940 in two technological sectors out of those selected for the recent period.*

Fall in concentration is identified with reference to a fall in the cross-firm coefficient of variation of the relevant index over time. Rising or stable concentration is identified as a rising or a constant coefficient of variation respectively over time.

Source: Cantwell and Andersen 1996

Taking the degree of concentration in the absolute size of activity as a point of comparison, in the interwar period the decline in concentration was somewhat more pronounced, extending to 36 sectors (as opposed to 27), while in recent period it was slightly less evident with 37 sectors showing a fall in concentration (as against 39).

The longer-term trend towards the deconcentration of technological activity is fairly clear; not only from the evidence of a very high degree of technological diffusion of each technology *within* both historical periods, but also *across* the two historical periods. In all the ten sectors that were fast growing in both 1920-40 and 1960-90, the degree of concentration of leadership was lower in 1990 than it had been in 1920. This fall in the concentration of leadership applies even when considering only historically established firms and not recent new entrants from the US and Europe (not to mention the firms of Japan and the Pacific Rim). There is also a general fall in concentration in the size distribution of activity, which applies to eight out of the same nine sectors (the exception being internal combustion engines). As mentioned earlier, this deconcentration is in accord with the technological diversification of corporate groups, as they have tended to move away from a narrower specialisation and towards broader systems for technological development.

Schumpeter (1934, 1939) argued that leading companies would pioneer a new area of technological development, but if successful they would be followed by a wider fringe of competitors which would imitate them and 'catch up'. At this stage the growth of the new technology would take off. On this basis a Schumpeterian hypothesis can be formulated. The measure of corporate concentration is expected to decline most (and leadership to disperse most) when the growth of the overall stock of patenting in a technological sector grows fastest. To this end, changes in the concentration of corporate technological leadership over five year periods were calculated and compared with the growth rate of accumulated patent stock for each of the selected patent classes.

When measuring innovation diffusion one of the theoretical internal (or endogenous) diffusion models described in Chapter 4, as for example, the epidemic model which has been applied by e.g. Mansfield (1961), would of course be appropriate to consider. However, the issue addressed here is not as stylised as these theoretical internal diffusion models often reveal. For example, as most corporate groups are patenting within a broad range of different technological fields, what is important is their *degree* of specialisation or focus within the different technological fields, rather than whether they specialise or patent at all within these technological fields. Hence, the change in the degree of concentration of technological leadership is in this case, and more appropriately, calculated as the average percentage change in the coefficient of variation over five-year time-periods within the two historical periods from 1920 to 1940 (from 1920-25 through to 1935-40) and from 1960 to 1990 (from 1960-65 through to 1985-90). As a measure of growth, the changes in the overall total accumulated patent stock for each of the selected patent classes are calculated in a similar way over the same periods.

*Table 8.4**. Testing an application of the Schumpeterian hypothesis

The five-year period with the highest rate of dispersion of corporate technological leadership (RTA) is also the period with the highest rate of technological growth (by numbers of sectors)						
	interwar (1920-40)			recent (1960-90)		
Technological group	Max. growth = max. diffusion	Max. growth lags one period	No relationship	Max. growth = max. diffusion	Max. growth lags one period	No relationship
Chemical	4	5_2	3	4_2	6_4	2
Electrical	7_2	2_1	3	8_5	2_1	2
Mechanical	7_3	1	4	5_1	2	5
Transport	3	0	3	4_1	2	0
The implication of size: The five-year period with the highest dispersion of the size distribution of corporate technological activity (PS) is also the period with the highest rate of technological growth.						
Chemical	3_1	1_1	8	7_1	2_1	3
Electrical	3_1	3_2	6	7_1	1_1	4
Mechanical	5_2	5_2	2	6_2	2_2	4
Transport	0	0	6	3	1_1	2

** The subscript refers to the number out of the box's total, where the hypothesis only fits with the second highest growth rate.*
Max. diffusion is identified as the 5-year interval in which there is the greatest fall in concentration, as measured by the largest fall in the coefficient of variation.
Max. growth is identified as the 5-year interval in which there is the largest percentage increase in accumulated patent stock.
Source: Cantwell and Andersen 1996

Table 8.4 presents some interesting results concerning this application or version of a Schumpeterian hypothesis. These results show that the periods in which the largest decline in concentration of technological leadership was experienced roughly correspond to the periods in which the growth in accumulated patent stock was highest in both the interwar and recent periods.

Out of 42 sectors there were 13 in which there was no relationship between the period of greatest corporate deconcentration and the period of the fastest growth in patenting in the interwar period, and just nine such sectors in the recent phase of development. The linkage between corporate deconcentration and the overall growth of activity holds equally for all four broad industrial groups (Chemicals, Electrical/electronics, Mechanical and Transport) in their most active technical fields. It is not surprising that the best results were obtained in the recent period, given that the accumulated patent stock for sectors and firms are on average much larger within that period, so small number problems are avoided to a greater extent. That the maximum rate of growth in accumulated patent stock often tends to lag the greatest dispersion of leadership by one period is consistent with the Schumpeterian hypothesis, which supposes that the takeoff in technological development is a result of the technology diffusion process. For example, it might be that growth tends to lag a period of technological experimentation combined with corporate technological competition.

Similar results were obtained for the cross-firm dispersion of the absolute magnitude of activity in a sector, when comparing the extent of coincidence between the growth of fields and the decline in concentration in the cross-firm distribution of the level of activity (using the PS index). However, the relationship here was rather weaker, with no linkage between growth and the inter-company dispersion of activity in 22 sectors in the interwar period (as opposed to 13), and 13 sectors in the recent period (as against 9). This additional evidence is not strictly relevant, though, to the Schumpeterian hypothesis that was formulated above. What is being measured through the PS index is different, and trends in the distribution of the PS index across firms therefore must be interpreted differently. As argued above, a decline in the cross-firm concentration of the index of technological specialisation (RTA) amounts to a process of diffusion, in that an increasingly wider range of firms orient their resources towards development in the field in question. In addition to this, a fall in the cross-company concentration of the size distribution of technological activity (the PS index) represents the movement of a wider range of firms into corporate research on a large scale in the field in question. This will be associated with faster growth to the extent that it is linked to diffusion in a particular field (the RTA and PS indices are correlated), or that fluctuations in growth are synchronised across fields, perhaps due to macroeconomic trends. (Chapter 6 also illustrated how such trends are related to firms and industrial sectors coming together in broader knowledge bases.)

8.5. THE EXTENT OF CONTINUITY IN CORPORATE TECHNOLOGICAL LEADERSHIP

The analysis of the changes in the degree of concentration of technological leadership can be further statistically decomposed into a 'regression' and 'mobility' effect (also applied in Chapter 3) in order to examine the extent of changes in the actual composition of corporate technological leadership.

For selected sectors of fast growing activity cross-firm regressions were run of the value of the index of corporate technological specialisation (RTA) at a given point in time on the equivalent value five years earlier. The five year periods selected were those in which each sector experienced the greatest diffusion of leadership or 'catching up' (as identified in Section 8.4). The correlation between the cross-firm distribution of the RTA index within a sector at time t and the earlier t-1 (where time is measured in five-year intervals) was estimated for each relevant technological sector through the following form of simple cross-section regression:

(iii) $$RTA_{it} = \alpha + \beta RTA_{it-1} + \varepsilon_{it}$$

The subscript i refers to firms of the pertinent industrial group active in the period in question. The standard assumption of this analysis is that the regression is linear and the residual ε_{it} is independent of RTA_{it-1}. This is valid if the cross-firm RTA index approximately conforms to a bivariate normal distribution, which requirement is met if there is an adequate number of patent counts in the sector as a whole and in the total patenting of the firms under consideration. Firms were included in the regression for each sector linked to their industrial group if they were active at all in patenting in the period in question, irrespective of whether they were active in that particular sector; hence, RTA values of zero were included where firms had been granted patents in other sectors, but not in the class being considered.

In this context the 'regression effect' is an indicator of whether the leading firms in the sector tend to slip back in terms of the intensity of their specialisation, or whether instead they tend to improve their position as a group. If, in the intermediate case, the relative specialisation of firms is unchanged, then $\beta = 1$; while, if followers tend to 'catch up' with the leaders in the sector then $\beta < 1$ (regression towards the mean). Hence the size of the regression effect will be measured by $(1 - \beta)$, and this is estimated by $(1 - \hat{\beta})$. A test of the strength of the regression effect is the t-test of whether $\hat{\beta}$ is significantly less than one, which amounts to a test of whether $(1 - \hat{\beta})$ is significantly greater than zero. Where corporate technological leadership is dispersed to the extent that followers catch up significantly, it is to be expected that the regression effect will be significant.

Table 8.5. Statistically testing the continuity of technological leadership of Chemical firms using cross-firm regression across interwar period: $[RTA]_{i(t)} = \alpha + \beta\ [RTA]_{i(t-1)} + \varepsilon_{(t)}$

		REGRESSION EFFECT		MOBILITY EFFECT		
Technological sector	Period	$(1-\hat{\beta})$	$t_{\beta 1}$	$(1-\hat{\rho})$	$t_{\beta 0}$	Number of active firms
CHEMICALS (INTERWAR)						82
62	1925-30	-0.102	1.181	0.102	12.767**	41
524	*1935-40*	*0.631*	*-14.093***	*0.245*	*8.224***	*53*
525	1920-25	-0.082	0.336	0.410	4.439**	39
526	1930-35	0.677	-4.373**	0.700	2.082*	46
528	1920-25	0.483	-6.738**	0.236	7.200**	39
536	1920-25	0.356	-24.004**	0.010	43.452**	39
558	1925-30	0.770	-9.638**	0.582	2.875**	41
568	*1935-40*	*-0.225*	*6.475***	*0.328*	*6.482***	*53*
570	*1920-25*	*0.644*	*-12.375***	*0.252*	*6.851***	*39*
514	1925-30	0.361	-4.579**	0.208	8.102**	41
208	1930-35	0.335	-6.735**	0.104	13.368**	46
585	1930-35	0.660	-5.506**	0.607	2.838**	46

Note: See Tables 7.4-7.11 in Chapter 7 (Appendix section 7.6.1) to match technological sector code with full description (name and technological group belonging).
*** : significant at the 1 % level; and *: significant at the 5 % level.*
Classes in which the Schumpeterian hypothesis was rejected are shown in italics.
Source: Cantwell and Andersen 1996

Meanwhile, the estimate of Pearson's correlation coefficient (ρ) between the two RTA distributions, $\hat{\rho}$, provides a measure of the mobility of the individual firms up and down the RTA distribution. A high value of ρ indicates that the relative strength of the specialisation of firms active in the sector is little changed, while a low value indicates that some firms are moving up and others are moving down the RTA distribution (having allowed for any general regression effect), quite likely to the extent that the ranking of firms in terms of their degree of specialisation within the sector changes. Thus, the magnitude of $(1 - \hat{\rho})$ measures the 'mobility' effect.

Table 8.6. Statistically testing the continuity of technological leadership of Chemical firms using cross-firm regression across recent period : $[RTA]_{i(t)} = \alpha + \beta\ [RTA]_{i(t-1)} + \varepsilon_{(t)}$

Technological sector	Period	REGRESSION EFFECT $(1 - \hat{\beta})$	$t_{\beta 1}$	MOBILITY EFFECT $(1 - \hat{\rho})$	$t_{\beta 0}$	Number of active firms
CHEMICALS (RECENT PERIOD)						82
71	1965-70	0.349	-4.696**	0.257	8.749**	64
210	1970-75	0.493	-11.215**	0.172	11.512**	63
430	*1985-90*	*0.178*	*-6.632***	*0.031*	*30.677***	*83*
512	1965-70	0.201	-1.775	0.336	6.987**	64
521	1970-75	0.351	-6.920**	1.147	12.778**	63
522	1965-70	0.347	-14.089**	0.041	26.536**	64
523	1965-70	0.430	-2.326*	0.635	3.083**	64
540	*1970-75*	*0.425*	*-11.652***	*0.104*	*15.773***	*63*
424	1965-70	0.660	-14.545**	0.310	7.499**	64
514	1965-70	0.186	-3.307**	0.122	14.469**	64
435	1965-70	0.321	-6.854**	0.121	14.494**	64
436	1960-65	0.542	-20.517**	0.097	17.317**	70

*** : significant at the 1 % level; and *: significant at the 5 % level.*
See further Table 8.5.

The stronger the correlation, the greater the persistence of leadership from one period to the next, and the weaker the mobility effect. If the mobility effect is low then existing leaders incrementally consolidate, or at least preserve, rather than lose their position as new technological fields emerge. We might expect this to happen over shorter periods of time (or within 'paradigms'), due to the strong stability which was found in Chapter 7, with respect to the position of firms, types of corporate technological competencies within the interwar or recent postwar periods, even though the composition of leadership may shift over longer periods of time or between 'paradigms' (comparing 1990 with 1940), as was also illustrated in Tables 8.1 and 8.2.

Table 8.7. Statistically testing the continuity of technological leadership of Electrical/electronic firms using cross-firm regression across interwar period: $[RTA]_{i(t)} = \alpha + \beta \ [RTA]_{i(t-1)} + \varepsilon_{(t)}$

		REGRESSION EFFECT		MOBILITY EFFECT		
Technological sector	Period	$(1-\hat{\beta})$	$t_{\beta 1}$	$(1-\hat{\rho})$	$t_{\beta 0}$	Number of active firms
ELECTRICAL/ELECTRONICS (INTERWAR)						45
332	1920-25	0.875	-8.588**	0.723	1.224	20
370	1930-35	0.079	-4.899**	0.004	57.215**	29
455	1920-25	0.617	-10.491**	0.162	6.523**	20
342	1920-25	0.613	-17.704**	0.065	11.194**	20
343	1920-25	0.567	-16.072**	0.055	12.293**	20
358	1925-30	0.783	-0.822	0.953	0.228	26
328	1920-25	0.478	-2.974**	0.392	3.245**	20
330	1935-40	0.492	-10.602**	0.106	10.954**	32
331	1930-35	0.249	-8.001**	0.022	24.139**	29
333	1925-30	0.264	-1.203	0.434	3.360**	26
62	1930-35	0.265	-3.655**	0.110	10.156**	29
357	1930-35	0.952	-10.556**	0.899	0.526	29

Note: See Tables 7.4-7.11 in Chapter 7 (Appendix section 7.6.1) to match technological sector code with full description (name and technological group belonging).
*** : significant at the 1 % level; and *: significant at the 5 % level.*
Classes in which the Schumpeterian hypothesis was rejected are shown in italics.
Source: Cantwell and Andersen 1996

An inverse test of the strength of the mobility effect is the t-test of whether $\hat{\beta}$ is significantly greater than zero, which in a simple bivariate regression is equivalent to the F-test on the significance of the correlation of the regression as a whole, and this amounts to a test of whether $\hat{\rho}$ is greater than zero. So, if

$\hat{\beta} > 0$ and $\hat{\rho} > 0$ it suggests that the mobility effect measured by $(1-\hat{\rho})$ is comparatively weak; while if $\hat{\beta}$ is not significantly different from zero this is associated with a significant mobility effect in which $(1-\hat{\rho})$ is not far below unity.

Table 8.8. Statistically testing the continuity of technological leadership of Electrical/electronic firms using cross-firm regression across recent period: $[RTA]_{i(t)} = \alpha + \beta\ [RTA]_{i(t-1)} + \varepsilon_{(t)}$

		REGRESSION EFFECT		MOBILITY EFFECT		
Technological sector	Period	$(1-\hat{\beta})$	$t_{\beta 1}$	$(1-\hat{\rho})$	$t_{\beta 0}$	Number of active firms
ELECTRICAL/ELECTRONICS (RECENT PERIOD)						45
375	1960-65	0.576	-7.688**	0.319	5.663**	39
341	*1970-75*	*0.592*	*-26.370***	*0.037*	*18.136***	*28*
382	*1965-70*	*0.697*	*-7.814***	*0.430*	*3.401***	*26*
372	1965-70	0.556	-3.171**	0.541	2.528*	26
437	1965-70	0.228	-2.783*	0.112	9.436**	26
307	1965-70	0.440	-4.457**	0.244	5.665**	26
357	1965-70	0.317	-2.805**	0.223	6.043**	26
360	1960-65	0.508	-8.762**	0.188	8.469**	39
364	1960-65	0.654	-26.214**	0.084	13.873**	39
365	1965-70	0.284	-2.401*	0.223	6.042**	26
371	1965-70	0.721	-12.169**	0.307	4.705**	26
355	1965-70	0.474	-16.984**	0.032	18.845**	26

*** : significant at the 1 % level; and *: significant at the 5 % level.*
See further Table 8.7.

The values of the measures and the results of these tests derived from the cross-firm regressions are reported in Tables 8.5 to 8.12. The general strength of the regression effect across sectors is perhaps not surprising, in the view of the way in which the periods were selected. By choosing the periods in which the extent of corporate deconcentration was at its strongest in each sector, a catching up effect is to be expected.

Table 8.9. Statistically testing the continuity of technological leadership of Mechanical firms using cross-firm regression across interwar period: $[RTA]_{i(t)} = \alpha + \beta \; [RTA]_{i(t-1)} + \varepsilon_{(t)}$

		REGRESSION EFFECT		MOBILITY EFFECT		
Technological sector	Period	$(1-\hat{\beta})$	$t_{\beta 1}$	$(1-\hat{\rho})$	$t_{\beta 0}$	Number of active firms
MECHANICAL (INTERWAR)						107
426	*1935-40*	*0.328*	*-31.953***	*0.009*	*65.394***	*85*
29	1925-30	0.548	-10.226**	0.269	8.442**	64
148	1935-40	0.585	-71.247**	0.016	50.617**	84
206	1920-25	0.226	-4.038**	0.118	13.843**	57
162	1930-35	0.647	-59.122**	0.033	32.207**	75
165	1920-25	0.701	-21.871**	0.217	9.331**	57
65	1925-30	0.699	-14.456**	0.379	6.233**	64
428	1930-35	0.628	-24.489**	0.139	14.487**	75
501	*1925-30*	*0.155*	*-2.141**	*0.172*	*11.624***	*64*
350	1930-35	0.227	-3.370**	0.198	11.465**	75
352	*1935-40*	*0.288*	*-3.842***	*0.276*	*9.501***	*84*
356	1930-35	0.691	-16.391**	0.348	7.345**	75

Note: See Tables 7.4-7.11 in Chapter 7 (Appendix section 7.6.1) to match technological sector code with full description (name and technological group belonging).
*** : significant at the 1 % level; and *: significant at the 5 % level.*
Classes in which the Schumpeterian hypothesis was rejected are shown in italics.
Source: Cantwell and Andersen 1996

Indeed, the regression effect was positive and statistically significant at the 5% level in 36 out of 42 sectors in the interwar period (in 34 at the 1% level), and in 38 out of 42 sectors in the recent period (in 34 at the 1% level).

However, in three of the remaining sectors in the interwar period - all in Chemicals - the regression effect was actually negative ($\hat{\beta}$ was greater than one). In other words, in these sectors the standard deviation of the cross-firm RTA distribution actually increased during the relevant five-year interval, but our measure of concentration (the coefficient of variation) fell as the mean of the RTA distribution rose by more. This was the result of starting with a very high degree of concentration and low mean value, which was partly attributable to low levels of patenting (these sectors being in the very early stages of corporate technological development). The three sectors are 568 (other organic compounds), 62 (refrigeration) and 525 (synthetic resins and fibres).

Table 8.10. Statistically testing the continuity of technological leadership of Mechanical firms using cross-firm regression across recent period: $[RTA]_{i(t)} = \alpha + \beta\ [RTA]_{i(t-1)} + \varepsilon_{(t)}$

Technological sector	Period	REGRESSION EFFECT $(1-\hat{\beta})$	$t_{\beta 1}$	MOBILITY EFFECT $(1-\hat{\rho})$	$t_{\beta 0}$	Number of active firms
MECHANICAL (RECENT PERIOD)						107
228	*1970-75*	*0.648*	*-64.986***	*0.031*	*35.211***	*82*
623	*1980-85*	*0.088*	*-2.575**	*0.051*	*26.556***	*80*
502	*1960-65*	*0.669*	*-8.632***	*0.593*	*4.274***	*94*
503	1970-75	0.503	-68.301**	0.009	67.517**	82
65	*1985-90*	*0.105*	*-6.135***	*0.014*	*52.406***	*80*
428	1970-75	0.519	-12.774**	0.201	11.879**	82
128	1985-90	0.590	-18.910**	0.170	13.145**	80
250	1965-70	0.498	-4.214**	0.577	4.251**	82
346	*1960-65*	*0.586*	*-32.859***	*0.076*	*23.212***	*94*
350	1970-75	0.072	-3.260**	0.022	41.989**	82
356	1965-70	0.054	-0.885	0.133	15.563**	82
604	1965-70	0.331	-4.654**	0.275	9.402**	82

*** : significant at the 1 % level; and *: significant at the 5 % level.*
See further Table 8.9.

Table 8.11. Statistically testing the continuity of technological leadership of Transport firms using cross-firm regression across interwar period: $[RTA]_{i(t)} = \alpha + \beta\ [RTA]_{i(t-1)} + \varepsilon_{(t)}$

		REGRESSION EFFECT		MOBILITY EFFECT		
Technological sector	Period	$(1-\hat{\beta})$	$t_{\beta 1}$	$(1-\hat{\rho})$	$t_{\beta 0}$	Number of active firms
TRANSPORT (INTERWAR)						50
123	*1935-40*	*0.252*	*-1.434*	0.364	4.874**	37
180	1920-25	0.655	-9.282**	0.294	4.889**	26
296	1920-25	0.773	-13.773**	0.363	4.045**	26
244	*1935-40*	*0.139*	*-2.246**	*0.086*	*13.915***	*40*
280	*1925-30*	*0.673*	*-6.307***	*0.512*	*3.060***	*32*
264	1920-25	0.744	-8.308**	0.495	2.863**	26

Note: See Tables 7.4-7.11 in Chapter 7 (Appendix Section 7.6.1) to match technological sector code with full description (name and technological group belonging).
*** : significant at the 1 % level; and *: significant at the 5 % level.*
Classes in which the Schumpeterian hypothesis was rejected are shown in italics.
Source: Cantwell and Andersen 1996

Despite the significance of the catching up phenomenon in the periods selected, there is little evidence of a shift in the ranking of firms over these reasonably short five year intervals. The mobility effect was significant ($\hat{\beta}$ was not significantly different from zero at the 5% level) in just three sectors out of 42 in the interwar years, and in none in the recent period. The three sectors in which there was a significant mobility effect at some point during the interwar period were 332 (telecommunication modulators), 358 (television equipment) and 357 (active solid state devices). However, a closer inspection of the actual shifts in the RTA distribution in these cases reveals that the significant mobility of firms is a reflection of the relatively low levels of corporate patenting in these sectors in the period selected. The low levels of corporate patenting in these cases created random fluctuations in the composition of technological leadership over even short periods of time.

Table 8.12. Statistically testing the continuity of technological leadership of Transport firms using cross-firm regression across recent period: $[RTA]_{i(t)} = \alpha + \beta\,[RTA]_{i(t-1)} + \varepsilon_{(t)}$

		REGRESSION EFFECT		MOBILITY EFFECT		
Technological sector	Period	$(1-\hat{\beta})$	$t_{\beta 1}$	$(1-\hat{\rho})$	$t_{\beta 0}$	Number of active firms
TRANSPORT (RECENT PERIOD)						50
123	1980-85	0.223	-8.912**	0.021	30.974**	43
180	1960-65	0.501	-11.173**	0.144	11.125**	47
296	1985-90	0.112	-1.767	0.091	13.997**	43
114	1970-75	0.469	-16.436**	0.054	18.629**	43
293	1970-75	0.1925	-0.945	0.473	3.965**	43
264	1960-65	0.408	-10.019**	0.092	14.557**	47

*** : significant at the 1 % level; and *: significant at the 5 % level.*
See further Table 8.11.

8.6. CONCLUSION

Few Schumpeterian studies have concerned themselves with the relationship between corporate technological leadership and innovation diffusion, despite the fact that this linkage lies at the heart of Schumpeter's theory of profits and growth. Indeed, most accounts seem to regard diffusion as a quite separate process. The investigation reported in this chapter represents a preliminary attempt to redress this incompleteness.

It was found that due to the historical shift towards more integrated technological systems through the fusion of diverse and formerly separate branches of technology, combined with companies' own growth, an overall historical trend of decline in the concentration of technological leadership and an increase in corporate technological diversification have occurred.

Furthermore, examining historical evidence of corporate patenting in the major international industries, it has been shown that there is a linkage between innovation diffusion among the largest firms and a rapid rate of technological development.

During these periods of very fast growth in technological activity followers tend to catch up with the leading pioneers in the area concerned (there is a positive and significant 'regression' effect). However, while the degree of leadership is changed, it is not eroded entirely, and established leaders still tend to preserve their position in a more moderate form (there is not a significant 'mobility' in the ranking of firms in a fast growing sector). Hence rapid technological development in a sector is associated with convergence of technological leadership within the sector in question. This evidence very much accords with a Schumpeterian story.

9. Conclusion: Technological Change and the Evolution of Corporate Innovation

Abstract

Drawing upon the empirical evidence provided in the previous chapters, this final chapter summarises and maps the processes by which the technological, innovative and competitive landscapes within industrialised societies have evolved to become increasingly wide-ranging and complex.

It is argued that structural changes and 'waves' of advance are associated with processes (of selection, development/diffusion and variety creation) which are of a more *general* kind, and that this general nature needs to be recognised separately from the inherited and *specific* nature of paradigms and trajectories. It is the latter which provides a framework for understanding the 'rules' which are guiding the specific characteristics underpinning the change and development processes. As mentioned previously, the notion of 'rules' is in this book used in a way resembling 'rules of the game' in North (1990).

The implications the overall results have upon the social sciences and the interpretation of history are discussed. Among several different perspectives, it is suggested that economics should have a place in the art of history, just as history already has a purpose within evolutionary economics.

9.1. TOWARDS A CONCLUSION

In its methodological design, the exploration of some central dynamic processes of technological change and corporate innovation has *not* been about confirming what we already know, but about using and challenging what we already know in order to gain further insight.

A procedure of this book has been to map, confront, and resolve several 'beliefs' in Schumpeterian approaches within evolutionary economics, concerning the dynamic elements of technological change and the evolution of corporate innovation. Evidence has been discussed in relation to the overall changing structure of US patenting records 1890-1990. Special focus has been on the patenting records of the leading American and European companies belonging to corporate groups in major international industries.

In this concluding chapter, both the empirical and theoretical contribution of the study of change and development, which this book has provided, will be highlighted. Processes and underpinning 'rules' will be the unit of analysis.

9.1.1. Chapter Outline

Accordingly, this final Chapter will map some of the empirical and theoretical results of the research programme in this book. First, a brief summary of the historical findings, with respect to technological change and corporate innovation 1890-1990 is presented. This is followed by a more comprehensive discussion of the stylised facts this books has provided, concerning the processes and the underpinning 'rules' by which industrialised societies have evolved. The implications the overall results have upon the social sciences will then be discussed, before some concluding remarks and perspectives close the book.

The great diversity of themes which have been addressed in this volume have drawn upon a huge mass of quite diverse (although distinct) discussions, controversies and literature. Individual chapters, set out above, have referred to these in some detail.

9.2. BRIEF HISTORICAL SUMMARY

The last century of technological opportunities (1890-1990) can be divided into two major technological regimes or paradigms, governing the operation by which path-dependent evolution has moved us from one, to another, and towards a third wave of technological and corporate development and expansion, normally associated with upswing periods.

The first technological regime or paradigm extend from the opening years of this century until 1940. The postwar period was characterised by technological diversification within broad technological groups, reflecting the formation of a structure of specialised engineering and science-based fields. However, the one that has followed through to recent times (in which the gap between the science-based technologies on the one hand, and the engineering-based technologies on the other hand has been widening less quickly) is expressive of an historical shift towards more integrated technological systems through the fusion of diverse and formerly separate branches of technology (Chapter 3). The new paradigm builds to a greater extent on complementarity and interrelatedness rather than more isolated individual channels of development.

We also saw how technological trajectories governed by those technological regimes have co-evolved with a historical trend across technological systems, in which previously distinct industrial sectors appear to be coming together in adopting, exploiting and developing technological systems. Evidence also illustrated how expansion of corporate technological competencies increasingly has been across industry boundaries, both within and across broad industrial groups. This is reflected in the industrial sectors' broadening their technological competencies (Chapter 6).

In this context, a historical trend of technological diffusion (identified in Chapter 8) was explicitly associated with an increasing number of sub-sectors across industries playing an important role in the takeoffs of three waves of innovation (normally associated with technological and economic upswings) in the twentieth century (Chapter 6). Those innovation waves were set into motion by different rates of change and rates of diffusion in both technological activity (Chapter 4) and corporate activity (Chapter 8) along interrelated technological growth cycles (identified in Chapters 4 and 5), as well as associated with structural changes in technological and innovative landscape, governing both technological and business scope (Chapters 5 and 6).

The overall interwar innovation wave gave scope to a system based on coal as a feedstock, chemical process technology and technology related to products such as rubber (especially for tyres) and dyes. Along with this Chemical system emerged electrical engineering, telecommunication, a narrow range of optics, and a broad range of illumination. Developments also took place within heavy metal engineering and mining, as well as in aircraft technology. The overall postwar innovation wave gave scope to a much more complex system with extended interrelated technological and industrial boundaries. The system was based on petrochemical feed-stocks, synthetic materials, and plastics with broad technological applications. Non-metallic mineral products, such as glass and advanced ceramics, also advanced with great new opportunities. All those developments were interrelated with radio systems (especially RADAR technology), office equipment and data processing, as well as new optics. It was an energy intensive system with major growth in power plants (including nuclear technology). Development of engines and fuel was also central, especially for motor vehicles. The empirical evidence from the structure of patenting also forecast an emerging system which moves us into the life sciences and the exploitation of biotechnology and micro-electronics (Chapter 5 and 6). The three waves of technological change and corporate innovation were found, in this volume, suitable to compare with the third, fourth and fifth Kondratiev waves (Kondratiev 1925).

Finally, evidence revealed how firms within industries practised quite different competencies, strategies or technological profiles in their participation and interaction with the evolving technological and innovative landscapes. This documents how firms' - even within industries - exploit, develop and catch up in increasingly complex spheres in rather different ways. In this context, evidence also revealed how firms technological competencies (associated with present or past strategy) change only very little historically, placing them in quite different (competitive) positions when the technological and innovative landscape changes (Chapter 7). Hence the scope (and limits to the scope) of the managerial discretion of corporate technological strategy is more complex than the one given by the specificity of an industry.

Overall, this brief summary of the empirical evidence in this volume certainly illustrates that the technological, innovative and competitive landscapes in which industrialised societies evolve have become increasingly interrelated and complex. The next section summarises the processes which have guided this evolution, as well as the rules underlying those, as identified in the previous chapters of this book.

9.3. STRUCTURAL TRANSFORMATION PROCESSES: THE REVEALED STYLISED FACTS

The results in this book, concerning the structural transformation processes with respect to the evolution of technologies and corporate innovation will now be presented. This is done in connection with a range of 'stylised facts' which have been revealed in relation to the evidence informing this book, i.e. the structure of patenting 1890-1990.

9.3.1. The Revealed Stylised Facts Concerning Technological Change: The Nature of Technological Expansion in Particular Sectors

Chapter 3 found that path-dependent evolution of technological opportunities (as opposed to great mobility across technologies fields over time), has increasingly been channelled into wider-ranging and more complex technological systems. These were offshoots of creative incremental technological development in a variety of areas. It was revealed how the evolution of knowledge embodied in new technological opportunities is characterised by 'creative, incremental, interconnected accumulation', as opposed to creative destruction, so that new technological systems with new opportunities built upon old ones (e.g. complement and extend them) rather than substitute them (see Stylised Fact I). This resembles the public good character of knowledge, concerning how knowledge bases can be infinitely exploited - how they are used or not used, but never destroyed. Of course, this does not mean that the activity or application (and/or the scope of the activity or application) in which the knowledge is embodied is not creatively destroyed (see Stylised Fact X).

Furthermore, the interconnected accumulation towards broader and more complex technological systems caused an evolution towards indistinct technological opportunities. Thus, the scientific and technological source sectors and diffusion sectors have become increasingly interrelated. This suggests an interactive process of development (as opposed to the linear model of a one-way transfer of science and technology into entrepreneurial practice). It also suggests that it might be problematic to talk about so-called 'key' inventions or innovations for a specific field, since it may be in minor improvements in other areas which make the so-called key invention possible.

The nature of evolving technological opportunities:

Stylised Fact I: The evolution of knowledge embodied in new opportunities is characterised by 'creative, incremental, interconnected accumulation', as opposed to creative destruction:

- *New technological systems build upon old ones rather than substitute for them.*
- *Technological systems have become broader and more complex: evolution towards interrelated (or indistinct) technological opportunities is witnessed.*

The formation of broader interrelated technological systems might be taken to imply that the growth of individual sectors has become more co-ordinated over time, so that technology develops in a synchronised fashion. Accordingly, it could be imagined that such processes of 'creative, incremental, interconnected accumulation' in technological opportunities would give rise to uniform patterns of developments.

However, evidence revealed how the stylised fact of 'creative, incremental, interconnected accumulation' in technological opportunities did *not* give rise to any uniform structure or synchronised movement. This is both in terms of (i) the Revealed Technological Trajectories (RTT) of technological opportunities (c.f. Chapter 3), as well as in terms of (ii) the trajectories of technological activity taking off at the technological frontier, which was indeed characterised by a diversity in the properties which influence their dynamics (c.f. Chapters 4 and 5):

(i) Concerning the evolving trajectories of technological opportunities, it was revealed how they do not follow uniform movements, and cannot be explained by aggregate measures, as they follow different patterns of development. However, some trajectories are more likely to be followed than others. This, in turn, confirms the instituted nature of technological development (Chapter 3). It also suggests that, although this book does not support the notion of key inventions or innovations (see above), it has appeared that when science and technology develop over a relatively long period of time, significant (key) trajectories can be traced or determined.

The nature of evolving trajectories of technological opportunities:

Stylised Fact II: Some trajectories are more likely to be followed than others:
- *Confirms the instituted nature of technological development.*

(ii) Concerning the evolving technological frontier, a variegated structure at the frontier at any point in time was revealed (Chapters 4 and 5): no synchronised movements at the frontier were witnessed, but instead we saw different periods of takeoff across technologies. Nevertheless, innovation takeoff tends to be clustered within certain broad periodical bands, suggesting a certain wave-like pattern of development (further elaboration on this point follow below, see Stylised Fact VI). Furthermore, within such bands, evidence showed no common accumulated socio-economic competence or impact, as well as unequal technological opportunity, across technological fields or innovation growth cycles taking off at the frontier, verifying an uneven type of technological system development.

The nature of the evolving technological frontier:

Stylised Fact III: Frontier movement indicates a band-like character of takeoffs in development of related technological fields.
- *Technological takeoffs are clustered within certain time bands, suggesting wave like patterns of development.*

Stylised Fact IV: Variegated structure at the frontier:
- *Unequal accumulated socio-economic competence or impact across technological fields.*
- *Unequal technological opportunity across technological fields.*

Hence, although technology develops over a broad complex technological front, some technologies within the broad system are in the forefront or at takeoff at certain times, and other technologies at other times. This evidence also confirms the view, that the broad technological front moves in patches in which some technologies lead at certain times and others get left behind, but catch up when they are needed, eventually, in order to make the broad front move (see Chapter 1, section 1.3). Overall, this also suggests how cumulative, incremental and path-dependent technological evolution tends to focus on certain fields or specific devices at certain times, and other areas at other times, within a broader front of interrelated technological systems.

Thus, evidence certainly suggests that both innovation cycles and the technological frontier cannot be understood from aggregate measures, and reveals how the nature, scope and overall configuration of each innovation wave differ.

In this context, *structural changes* in the technological and innovative landscape were set into motion by the uneven opportunities, and uneven socio-economic importance *across* technological fields, catching up and cumulating at uneven rates (governed by the paradigm in question) to enhance economies of scope. Thus, emergence of new technological systems is not only about the activity of catching up and cumulating, but about the diversity of technology dynamics.

The nature of structural change and emergence: Technological perspectives:

Stylised Fact V: Uneven opportunities, and uneven socio-economic importance across technological fields, catching up and cumulating at uneven rates (governed by the paradigm in question) to enhance economies of scope, causes structural changes.

- *Emergence of new technological systems is not only about the activity of catching up and cumulating, but about the diversity of technology dynamics.*

However, the innovation *waves* were established due to the nature by which the individual technological fields mature and diffuse. That is, both the overall quantity of innovative activity *within* a technological sector, as well as the expansion, distribution and structure of corporate activity in that sector is not constant.

It was established that innovative activity tends to develop in *S*-shaped growth paths, constituted from increasing technological possibilities due to its own growth, together with an overall decreasing trend (or retardation process) of new technological potential (Chapter 4). It was also established how this is associated with the diffusion paths of corporate activity, given that expansion of corporate activity in a technological field is positively associated with the growth of that field (i.e. a high rate of change is associated with a high rate of diffusion, Chapter 8). It is this nature of evolution of technological and corporate innovation which has also been associated with bandwagon effects of technological activity (Schmookler 1966), or associated with 'entrepreneurial swarming' of corporate activity (Schumpeter 1939). The latter issue concerning the nature of technological diffusion is dealt with in section 9.3.2 (Stylised Facts XV, XVI).

Hence, Chapter 4 supported the now 'almost assumed' *S*-shaped metaphor or image of the way technology develops. In this context, it was also revealed that growth cycles are not subject to significant external shocks, as most cycles survive all the way to their long-run potential. Thus, as stated by Schumpeter (1939: chapters 1 and 2), although external factors (random historical events) account for much in economic fluctuations, it is still important to understand the mechanics and dynamics internal to the cyclical system. By such, no fixed cycle duration was observed either (Chapter 4).

The nature of innovation waves:

Stylised Fact VI: Innovation waves were established as the overall quantity of innovative activity within a technological sector, and the structure and distribution of corporate activity in that sector (see Stylised Facts XV, XVI) is not constant.

The nature of trajectories of technological activity:

Stylised Fact VII: The use of the now 'almost assumed' S-shaped image or metaphor of the way technology develops is justifiable:

- *Technological evolution tends to follow an S-shaped growth path.*

Stylised Fact VIII: Technological growth cycles are generally not subject to significant external shocks.

Stylised Fact IX: No fixed technological growth cycle duration.

Overall, the research findings suggest that the evolution of technology (c.f. trajectories of technological opportunities, trajectories of innovation growth cycles, and the secular movements (c.f. wave-like patterns) at the technological frontier) cannot be understood from aggregate measures. Rather, the diverse patterns of technology dynamics, accommodating the evolving technological landscape in which business opportunities can be exploited, is more useful from a system perspective.

The nature of technological evolution:

Stylised Fact X: Technological evolution can be characterised by 'radical incrementalism' or 'revolutionary gradualism'.

- *Although technologies in new systems build upon old ones, the technological scope and configuration of the new systems still substitute (or creatively destroy) old ones.*

Combining cyclical theories of technological evolution and the theories of path-dependency with a system perspective, this book has exposed how overall technological evolution can be characterised by 'radical incrementalism' or 'revolutionary gradualism (using the terminology of Metcalfe 1998)'. In this setting, technologies in new systems builds upon old ones, but the technological scope and overall configuration (i.e. overall boundaries and internal structures) of new systems still substitute (or creatively destroy) old ones.

9.3.2. The Revealed Stylised Facts Concerning Corporate Innovation: The Implications of the Broader Front of Society's Technological Base for Industrial Societies and their Firms

When investigating the evolution of corporate innovation, the book builds upon the notion that a central characteristic of technological change is the way in which it gives rise to business and technological opportunities, sometimes in several industrial sectors. However, the ability and determination to exploit those opportunities is associated with socio-economic and corporate competence, which is location or firm-specific, cumulative in character and often tacit. It is also related to the extent to which the technological and business opportunities concerned are appropriate to strategies of the different socio-economies and industrial sectors.

Chapter 6 illustrated how industrial sectors are relatively more inclined to patent in the technological areas that are most closely associated with their industry (i.e. Chemical industrial sectors patent in Chemical technologies, Electrical industrial sectors in electrical technologies, and so forth). However, it has been less so over time, as industrial sectors appear to be coming together in adopting and developing technologies of increasingly more complex systems. This is also reflected in the industrial sectors' widening their technological competencies. In this context, Chapter 6 indeed illustrated how increased technological interrelatedness is combined with a widening and increased complexity in corporate and industrial competencies, which have a tendency to make industry boundaries less well-defined, change the nature of the competitive environment, which in turn have some feedback effect on subsequent innovations and hence on the technological landscape. It is therefore argued that, conceptually, an 'industry system' and its closest competitive environment can only be defined in non-static terms, as both the technological landscape in which it operates and its competencies change over time. This also has implications for the limitations of the standard industrial classification scheme when it comes to understanding the competitiveness of firms only in relation to its product markets, rather than in relation to the technologies it uses to develop those. Hence, if an industry needs to be understood in relation to its closest long-run dynamic competitive environments the standard industrial classification scheme (based upon product markets) simply provides us with very little insight, as (i) firms' dynamic specialisation (or innovative position, defined by its technological specialisation) do not necessarily correspond to their counterpart product specialisation, and as (ii) the innovative and competitive landscape underpinning the dynamics of industry is constantly changing. The competence bloc approach applied in Chapter 6 concerning the analysis of technological dynamics was very useful in illustrating how a technological system approach made a difference in the way we understand the boundaries of an industry.

The nature of evolving corporate technological specialisation:

Stylised Fact XI: Industrial sectors are relatively more inclined to patent in the technological areas that are most closely associated with their industry, although this is less so over time.

- *A historical trend can be identified, revealing how industrial sectors are coming together in adopting and developing new technologies of new technological systems.*
- *A historical trend can be identified, revealing how industrial sectors continuously keep widening their technological competencies, beyond their historical core technological counterparts (i.e. Chemical industrial sectors move into electrical, mechanical and transport technologies, and so forth).*

Stylised Fact XII: Industry boundaries become less well-defined

Stylised Fact XIII: The innovative and competitive landscape underpinning the dynamics of firms within industries is constantly changing.

These Stylised Facts indicating a blurring of industry boundaries are consistent with the historical shift towards more integrated technological systems through the fusion of diverse and formerly separate branches of technology, which was identified in section 9.3.1. Furthermore, the historical trend of industrial sectors coming together in adopting and developing new technologies of new technological systems, while widening their technological competencies suggests a distributed nature of how systems evolve (c.f. a type of a distributed innovation process, as suggested by Andersen, Metcalfe and Tether 2000, Coombs and Metcalfe 1998). Alternatively, it may be a reflection on how firms are building their own networks of both intra- and inter-organisational linkages and relationships.

In Chapter 7 we saw how the direction by which firms have broadened their ability to exploit the changing technological structures has differed, despite a movement toward more interrelated and complex systems. This might be related to diversity of firms' technological strategies in which some firms pursue (i) technological leadership (i.e. seeking a relative high technological competence in the areas of greatest technological opportunities), (ii) core participation (i.e. seeking a relative high technological competence in core technologies of its industry), and/or (iii) technological diversification (i.e. seeking a relatively wide or broad spectrum of technological competencies). Alternatively, a reason for the diverse behaviour of firms could be related to the inherited routines and practices within firms.

Hence, with reference to some of the principal determinants of corporate competitiveness, a firm taxonomy was developed, in which the variety of firms became evident (Chapter 7). Grouping all firms within industries in relation to their types of technological profiles, the book also examined how the technological position of firms, in relation to the socio-economic paradigm in question, influenced their subsequent technological trajectories. It was illustrated how the technological competencies of firms are relatively stable, and do not tend to fluctuate (Chapter 7).

However, when technological regimes change, firms' technological positions or profiles generally become eroded. That is, when the business opportunities change across technological paradigms, new types and compositions of technological competencies and new strategies are required to sustain the corporate profile or position, but firms find it difficult to adjust. This explains how the relative competitiveness among firms shifts across paradigms and how new firms get favoured, and how the earlier 'best performers' tend to be 'locked into' the 'old' paradigm. These results confirm the incremental, cumulative and path-dependent nature of the trajectories of corporate technological competencies.

The nature of the evolution of corporate technological profiles:

Stylised Fact XIV: The technological competencies of firms are relatively stable, and do not tend to fluctuate greatly over time.

- *Corporate technological competencies change only incrementally, and accumulate along certain restricted trajectories of development.*
- *When technological regimes change, the relative position (or profile) of firms' dynamic capabilities (and associated competitiveness) become eroded.*

Examining historical evidence of corporate patenting in major international industries, it was revealed that there is a link between innovation diffusion among the largest leading firms and a rapid rate of technological development (Chapter 8). That is, during periods of very fast growth in technological activity in a sector, corporate technological followers tend to catch up with the leading pioneers in that technological sector. Hence rapid technological development in a sector is associated with convergence of technological leadership within the sector in question. This is very much consistent with a Schumpeterian story. However, while the degree of leadership is changed, it is not eroded entirely, and established leaders still tend to preserve their position in a more moderate form.

The nature of the evolution of corporate technological leadership:

Stylised Fact XV: Corporate technological leadership within a sector tends to diffuse in the course of the technological growth of that sector.

- *Rapid technological development in a sector is associated with convergence of technological leadership within the sector.*

Stylised Fact XVI: Diffusion of corporate technological leadership is associated with 'restricted' catching up, and very little mobility in the leadership ranking across firms.

- *Early movers within a trajectory set the characteristics of the evolutionary paths.*

Stylised Fact XVII: Combined with an increase in companies' technological size, and an increased technological corporate diversification, an overall historical trend of decline in the concentration of corporate technological leadership has been revealed.

This also confirms the assumption in Chapter 6, that by focusing on the firms within industrial sectors participating (or having relative high competencies) at an early stage in a technological growth curve (or trajectory) associated with the emergence of a new technological system, we capture the industrial sectors launching the system and thereby its characteristics.

Combined with an increase in companies' technological size, on average, Chapter 8 also revealed an overall historical trend of decline in the concentration of technological leadership, along with an historical trend of increase in corporate technological diversification.

With respect to structural changes (Stylised Facts XVIII, XIX) and wave-like patterns in the evolution of corporate innovation (Stylised Fact XX), the emergence, transformation, growth, and decline of industrial systems have co-evolved with the changing technological structures and innovation-waves, which were set out in the previous section (Stylised Fact V and VI).

That is, changes in the composition, boundaries, opportunities and relative importance of a structure (or system) of technologies do not happen in isolation, but in a dynamic interaction with its surrounding environment, including the industrial sectors and firms which act as the entities producing, modifying and using the component technologies (see Chapter 6). Just as a pervasive system selects the best-adaptable environment (including firms and industries) for diffusion, the firms and industries in that environment also contribute to, and thereby select, the rate and direction of technological activity. In this context, the interaction of firms with a changing structure of technologies can have sweeping consequences for the structure of firms' and industrial sectors' competencies, which in turn may change the nature of the competitive environment.

***The nature of structural change and emergence: industrial perspectives*:**

Stylised Fact XVIII: When technological structures change, this spills over to structural changes of industrial technological profiles (or positions), and vice versa.
Just as a pervasive technological system selects the best-adaptable environment (including firms and industries) for diffusion, the firms and industries in that environment also contribute to, and thereby select, the rate and direction of technological activity, and so forth.

Stylised Fact XIX: Uneven rates and directions of catching up and accumulation across firms, to enhance economies of scope within a given paradigm, causes structural changes.

Hence, structural changes of industries happen due both to (i) different rate and direction of corporate innovation (as well as catching up and accumulation), which evolve only incremental, path-dependent, and cumulative, as well as (ii) their co-evolution with changing technological regimes (as identified in previous section 9.3.1), which direct the scopes (and thereby the positions) for firms and industries (see also Stylised Fact XIV).

Along the same lines, the nature of waves of corporate activity, is also interrelated with nature of the technological growth curve (see Stylised Fact VII), in which Chapter 8 established how a rapid technological development in a sector is associated with diffusion of corporate activity within the sector (Stylised Facts XV). That is, a high rate of change is associated with high rate of diffusion.

(As mentioned in text to Stylised Fact VI (section 9.3.1) it is this nature and co-evolution of technology and corporate innovation which has been associated with secular movements of 'entrepreneurial swarming' (Schumpeter 1939) in relation to waves of corporate activity.)

Waves of corporate technological activity:

Stylised Fact XX: Waves of corporate activity were established as the overall quantity of innovative activity within a technological sector (Stylised Fact VII), and the structure or distribution of corporate activity in that sector (Stylised Fact XV), is not constant. (See also Stylised Fact VI.)

Overall, the nature of evolution of corporate and industrial technological scopes can (as technological evolution, see Stylised Fact X) also be characterised by 'radical incrementalism' or 'revolutionary gradualism'. That is, (i) although corporate innovation (including catching up and accumulation) evolve within technological structures, and *vice versa* (Stylised Fact XVIII), and (ii) although technologies and new systems built upon old ones (as identified in section 9.3.1), the corporate and industrial scopes governed by new technological systems still substitute (or creatively destroy) old ones.

The nature of the evolution of corporate and industrial technological scopes:

Stylised Fact XXI: The evolution of corporate innovation within industrial systems can be characterised by 'radical incrementalism' or 'revolutionary gradualism'.

- *Although corporate innovation (including catching up and accumulation) evolve within technological structures, and vice versa, and although technologies and new systems built upon old ones, the corporate and industrial scopes governed by new technological systems still substitute (or creatively destroy) old ones.*

9.3.3. Structural Transformation Processes: More Insight Needed

Of course there are other factors influencing structural transformation processes with respect to technological change and the evolution of corporate innovation, and some of those most important ones will now briefly be discussed. This is related to perspectives for future research and beyond the scope of this book.

Processes addressed within this book have primarily been in relation to how different rates of development across technologies and firms explain structural changes at the frontier, which sometimes evolves in waves. However, in this framework research within evolutionary economics also need to shed light on what triggers movements at the frontier, as well as the principles of the sustainability and performance of a socio-economic system or industrial society.

Empirical studies have also been partial, mainly focused on the behaviour of firms within individual industries rather than the inter-play between industries or industrial behaviour. This book certainly raised an issue with respect to the blurring of industry boundaries historically, and the need for integrating such perspectives into theoretical reasoning if firms are to be analysed and compared in relation to their closest competitors.

Also, transformations in industrial societies are not *only* caused (or naturally selected) by changing technological structures co-evolving with the development of corporate capabilities of industries (i.e. the endowments and resource-bases that are developed internally within firms, or distributed in cooperative arrangements among firms and other institutions), which have been the focus of this book, as in much evolutionary economics. More understanding is needed with respect to the interactive role and the dynamics of the demand side. We need to investigate the processes concerning how both the supply and demand conditions interact with the competitive and cooperative processes established in the macro environment or the market place.

Furthermore, we know that markets, generally, are not neutral, and often power structures and social constraints are influential factors, and this is also an area for future investigation. Also, on the supply side evolutionary economics has not made important contributions to the analysis of strategy or other related factors at the micro-firm level, which remain inside the black box (and only reflected in the corporate trajectories of firms, as is also the case of this book). That is, corporate strategy has, just as many technological phenomena, been treated as events transpiring inside a black box. (C.f. Rosenberg's (1982) 'black box' notion).

Also, empirical work within evolutionary economics has been rather restricted, mainly focused on manufacturing industries and technical innovations (important as they may be), overlooking services and non-technical innovations (e.g. within organisation or management).

Of course, there are other obvious factors not specified in the analysis of this book – such as the public and private R&D financing structure, the commercial R&D system, raw materials and energy sources, regulation, marketing etc. – which may also influence both corporate and technological behaviour, and thus the overall evolution of a socio-economic system.

9.4. CONCLUDING REMARKS

This volume has been concerned with investigating many of the factors and characteristics that are important for understanding the processes of change and development and the underpinning 'rules'. In this context, this book has provided some definitive treatment of the building blocks associated with dynamic processes within evolutionary economics (set out in Chapter 1) which so far have been treated very informally. Focus in this volume has been on technological change and the evolution of corporate innovation. The study has included a diversity of themes relevant to economists, historians, institutionalists, business-scholars and business-cycle theorists.

This treatment of dynamic processes has also provided some insight into how to analytically address a lot of concepts such as instituted economic processes, path-dependency, competition etc. Writing this book did not only open up for how informally many of those concepts are treated within evolutionary economics, but also for how important an understanding of such are for how we measure and understand the evolution of a socio-economic system.

9.4.1. Processes and Underpinning 'Rules'

It was revealed how instituted technological change, governed by major technological epochs or regimes of waves of development and structural change, have become increasingly interrelated across technological fields, and how path-dependent evolution of corporate innovation has become increasingly manifold in the knowledge bases upon which it draws. Hence, the technological, innovative and competitive landscapes within industrialised societies have become increasingly complex historically.

However, it is here argued that it is not the processes of co-evolving technological and industrial structures or the nature of the waves which have caused increased complexity historically, but that it is the 'rules' underpinning the processes, governed by the paradigms, which have resulted in such trajectories. Having applied the building blocks within evolutionary economics theoretical reasoning (which were set out in Chapter 1, section 1.3) throughout this book, we are now able to distinguish processes from 'rules', when understanding change and development.

Processes: With respect to understanding economic processes, the book has revealed how different rates of development across technologies and firms creates (or selects) new technological and industrial structures. Together with different rates of innovation diffusion along interrelated innovation trajectories or cycles, this spurs takeoffs of new waves of development at the technological frontier. This, in its turn, transforms the innovative and competitive landscapes, in open-ended processes of continuous change and development.

'Rules': With respect to the 'rules' underpinning the dynamic processes, the book has revealed how the technological paradigms stimulate or constrain the rates and directions of the evolution taking place. Hence, the paradigm provides a framework for how economic processes become instituted in systems and trajectories.

When analysing the 'rules' underpinning dynamic processes in an endogenous fashion, focus was on *deriving* trajectories and systems of various kinds (rather than assuming their existence). The purpose was to understand how populations (and their internal configuration) of technological sectors, firms and structure of industries have evolved and differed, and how changes occur only in a restricted fashion.[1] It was revealed how this has implications for how we need to understand any socio-economy (or industrialised society) and competitive environment in a flexible manner. Thus, the approach concerning deriving trajectories and systems of various kinds seems to be of particular importance when understanding the role of history and context specificity.

Moreover, deriving technological systems and trajectories was indeed useful in capturing structures, detail, and variety of historical changes, as well as overview. It allowed us to avoid the analytical tradeoff between complexity and overview, or between the long term and the detail, as it is able to provide both simultaneously. In its turn, this book has illustrated how a system and trajectory perspective seem to be a very appropriate 'analytical' tool (and not just a descriptive framework) when aiming to map the structural elements of innovation dynamics. Evidence also revealed how the 'rules' within evolutionary economics so far has been treated as very approximate, and that

[1] For elaboration, see also Chapter 1, section 1.3 on Building Blocks for Dynamic Processes.

only by applying them, do they become operational. As the specific meaning of the 'rules' cannot be understood without application and operation, this emphasises the importance of empirical research of this kind.

Hence, based upon various puzzles of conceptualising and measuring in this book, it is here proposed that structural changes and 'waves' of advance is associated with processes (of selection, development/diffusion and variety creation) whose nature are of more general kind. This general nature needs to be recognised separately from the inherited and specific nature of paradigms and trajectories. It is the latter which provides a framework for understanding the 'rules' which guides the specific characteristics underpinning the change and development processes.

Processes and underpinning 'rules': What is specific and what is general:

Structural changes and 'waves' of advance are associated with processes (of selection, development/diffusion and variety creation) whose nature are of more general kind, and this general nature needs to be recognised separately from the inherited and specific nature of paradigms and trajectories. It is the latter which provides a framework for understanding the 'rules' which guides the specific characteristics underpinning the change and development processes.

What is specific and what is general in the nature by which processes and 'rules' form and govern is indeed important when aiming to understand the performances of specific socio-economic systems or specific industrial societies and why they differ.

Since the 'rules' guide the procedures (stimuli and constraint) of the dynamic processes, by introducing institutional and historical aspects into the story, more variables become endogenous in the theoretical reasoning than in mainstream economics. Furthermore, as the character of the 'rules' are also a consequence of the outcome of its own procedures, the 'rules' also become a guidepost for change. Hence, almost 'everything' now has become endogenous in the evolution of a dynamic system, so an issue to be raised is of course to what extent economists should search for a full endogenous understanding of historical regularities in the processes of change and development, or to what extent economists should make allowance for random events or treat everything exclusively. It is well known that, whereas economists are more concerned with 'mechanisms' and processes, historians and business historians are more concerned with the 'flesh on the bone'. It is here suggested that by distinguishing between the general nature of processes and the 'rules' (i.e. the specific character underpinning the processes) any such tradeoff is minimised.

Moreover, in such a way economics has even found its place in the art of history, just as history already has a purpose within evolutionary economics.

In any case, it is evident that the way we measure, conceptualise and theorise change and development determines how we understand the innovative and competitive scenery that surrounds us. It is hoped this volume has some implications for how we decide to treat many dynamic variables within evolutionary economics.

References

Achilladelis, Basil, Schwarzkopf, Albert, and Cines, Martin (1987): 'A Study of Innovation in the Pesticide Industry: Analysis of the Innovation Record of an Industrial Sector', *Research Policy*, vol. 16. pp.175-212.

Andersen, Birgitte (1997): 'A Century of Technological Opportunities using Patent Data as a Source of Interpretation 1890-1990', Andersen, Birgitte: *Technological Change and The Evolution of Corporate Innovation.* PhD-thesis, Department of Economics, University of Reading.

Andersen, Birgitte (1998): "The Evolution of Technological Trajectories 1890-1990", *Structural Change and Economic Dynamics*. Vol. 9 No.1. pp.5-35.

Andersen, Birgitte (1999a): 'The Hunt for *S*-shaped Growth Paths in Technological Innovation: A Patent Study'. *Journal of Evolutionary Economics* (9): pp.487-526.

Andersen, Birgitte (1999b): 'The Complexity of Technology Dynamics: Mapping Stylised Facts in Post-Schumpeterian Approaches with Evidence from Patenting in Chemicals 1890-1990', in Louca, Francisco (ed.): *Perspectives on Complexity within Economics*, UECE (ISEG), pp.111-142.

Andersen, Birgitte and Cantwell, John A. (1999): 'How firms differ in their technological competencies and how it matters', *CRIC Discussion Paper* No 25.

Andersen, Birgitte and Howells, Jeremy (2000): 'Intellectual Property Rights Shaping Innovation Dynamics in Services', in: Birgitte Andersen, Jeremy Howells, Richard Hull, Ian Miles, and Joanne Roberts (eds): *Knowledge and Innovation in the New Service Economy*, Cheltenham: Edward Elgar Publ. pp.141-162.

Andersen, Birgitte and Lundvall, Bengt-Åke (forthcoming): 'Industrial Dynamics', in Miche, Jonathan (ed.), *Reader's Guide to the Social Sciences*, Vol.I and II, London: Fitzroy Dearnorn Publishers.

Andersen, Birgitte, Metcalfe, J. Stanley, Tether, Bruce S. (2000): 'Distributed Innovation Systems and Instituted Economic Processes', in J. Stanley Metcalfe and Ian Miles (eds), *Innovation Systems in the Service Economy: Measurement and Case Study Analysis*, Dordrecht: Kluwer: pp.15-42.

Andersen, Birgitte and Walsh, Vivien (2000): 'Co-Evolution With Chemical Technological Systems: A Competence Bloc Approach'. *Industry and Innovation*, Vol.7 No.1. pp.77-115.

Andersen, Esben Sloth (1994): *Evolutionary Economics. Post-Schumpeterian Contributions*. London: Pinter Publishers.

Archibugi, Daniele (1992): 'Patenting as An Indicator of Technological Innovation: A Review'. *Science and Public Policy*, Vol.19, No.6. pp.357-368.

Bailey, Kenneth D. (1994): *Typologies and Taxonomies. An Introduction to Classification Techniques,* Series: Quantitative Applications in The Social Sciences. Thousand Oaks: Sage Publications.

Bain, Andrew D. (1964): *The Growth of Television Ownership in The United Kingdom Since The War. A Lognormal Model*. Cambridge University Press: Cambridge.

Beggs. J. J. (1984): 'Long-Run Trends in Patenting'. Zvi Griliches, (ed.), *R&D. Patents and Productivity*,Chicago: University of Chicago Press. pp.155-173.

Brouwer, Maria (1991): *Schumpeterian Puzzles: Technological Competition and Economic Evolution*, Hemel Hempstead: Harvester Wheatsheaf.

Burns, Arthur F. (1934): *Production Trends in the US since 1870*, NBER and Kelly.

Cantwell, John A. (1991): 'Historical Trends in International Patterns of Technological Innovation', in James Foreman-Peck (ed.): *New Perspectives on The Late Victorian Economy: Essays in Quantitative Economic History, 1860-1914*. Cambridge University Press: Cambridge. pp.37-72.

Cantwell, John A. (forthcoming): *The History of Technological Development in Europe and the United States*, Oxford: Oxford University Press.

Cantwell, John A. and Andersen, Birgitte: (1996). 'A Statistical Analysis of Corporate Technological Leadership Historically', *Economics of Innovation and New Technology*, Vol. 4, No.3. pp.211-234.

Cantwell, John A. and Bachmann, Astrid (1998), 'Changing patterns in technological leadership - evidence from the pharmaceutical industry', *International Journal of Technology Management*, Vol. 21, No. 1, pp. 45-77.

Cantwell, John A. and Fai, Felicia (1999): 'The changing nature of corporate technological diversification and the importance of organisational capability' in Peter E. Earl and Sheila Dow (eds), *Contingency, Complexity and the Theory of the Firm. Essays in the Honour of Brian J. Loasby*, Vol. 2. Cheltenham, Edward Elgar. pp.113-137.

Cantwell, John A. and Sanna-Randaccio, Francesca (1993): 'Multinationality and Firm Growth', *Weltwirtschaftliches Archiv*, Vol. 129, No.2. pp.275-299.

Cantwell, John A. and Santangelo, Gracia D. (2000), 'Capitalism, innovation and profits in the new techno-economic paradigm', *Journal of Evolutionary Economics*, Vol. 10, Nos. 1-2, pp. 131-157.

Carlsson, Bo (1987), 'Reflections on Industrial Dynamics: The Challenges Ahead, International', *Journal of Industrial Organisation*, 5, pp.135-148.

Carlsson, Bo (ed.) (1995) 'Technological Systems and Economic Performance: the Case of Factory Automation', Boston / Dordrecht / London: Kluwer Academic Publishers.

Carlsson, Bo (ed.) (1997): *Technological Systems and Industrial Dynamics*, Mass: Kluwer Academic Publishers.

Carlsson, Bo (1998): 'Industry Clusters: Biotechnology/Biomedicine and Polymers in Ohio and Sweden', *CRIC Systems and Services Workshop*, March.

Carlsson, Bo, Eliasson, Gunnar and Taymaz, Erol (1997): 'The macro-economic effect of technological systems; Micro-Marco simulation', Carlsson, Bo (1997): *Technological Systems and Industrial Dynamics*, Mass: Kluwer Academic Publishers.

Carlsson, Bo and Stankiewicz, Rickard (1991): 'On The Nature, Function and Composition of Technological Systems', *Journal of Evolutionary Economics*, Vol. 1 No.2, pp. 93-118.

Chandler, Alfred D. Jr. (1990): *Scale and Scope: The Dynamics of Industrial Capitalism*, Cambridge, Mass.: Harvard University Press.

Cheung, Steven N. S. (1986) 'Property rights and invention' in J. Palmer (ed.): *Research, in Law and Economics: The Economics of Patents and Copyrights*, pp.5-18.

Chow, Gregory C. (1967): 'Technological Change and The Demand for Computers', *American Economic Review*, Vol.57. pp.1117-1130

Clark, James B. (1907) *Essentials of Economic Theory,* Macmillan, New York.

Coombs, Rod and Metcalfe, J. Stanley (1998): 'Distributed Capabilities and the Governance of the Firm', *CRIC Discussion Paper series*, No.16.

Coombs, Rod, Richards, Albert, Saviotti, Paolo, and Walsh, Vivien (1996): Technological Collaboration and Networks of Alliances in the Innovation Process'. Editors' Introduction to *Technological Collaboration: The Dynamics of Cooperation in Industrial Innovation*. Cheltenham: Edward Elgar.

Coombs, Rod, Saviotti, Paolo and Walsh, Vivien (1987): *Economics and Technical Change*, Macmillan, Basingstoke and London.

Cordada, James (1993): *Before the Computer. IBM, NCR, Burroughs, and Remington Rand, and The Industry they created, 1865-1965,* Princeton: Princeton University Press.

Dahmen, E. (1989): 'Development Blocks in Industrial Economics', in Carlsson, Bo (ed.) *Industrial Dynamics*, Kluwer Academic Publishers: Boston.

Day, Lance (1990): 'The Chemical and Allied Industries', in Ian McNeil (ed.): *An Encyclopaedia of the History of Technology*, London: Routledge. pp.186-227.

Dosi, Giovanni (1982): 'Technological Paradigms and Technological Trajectories: A Suggested Interpretation of the Determinants and Directions of Technical Change', *Research Policy*, vol. 11, No.3. pp.47-162.

Dosi, Giovanni (1988): 'The Nature of the Innovation Process', in Giovanni Dosi ; Christpoher Freeman; Richard Nelson; Gerald Silverberg; and Luc L.G. Soete (eds.): *Technical Change and Economic Theory*, London: Pinter Publishers. pp.221-238.

Dosi, Giovanni, Freeman, Chris, Nelson, Richard. R., Silverberg, Gerald and Soete, Luc.L.G. (eds.) (1988): *Technical Change and Economic Theory*, London: Pinter Publishers.

Du Boff, Richard B. (1967): 'The Introduction of Electric Power in American Manufacturing', *Economic History Review*, Vol.20. pp. 509-518.

Edgerton, David E.H. (1988): 'Industrial Research in the British Photographic Industry, 1879-1939', Jonathan Liebenau (ed*.), The Challenge of New Technology. Innovation in British Business Since 1850*. Gower: Aldershot, pp.106-134.

Edgerton, David E.H. (1991): *England and the Aeroplane: An Essay on a Militant and Technological Nation.* Basingstoke: Macmillian Press in association with the Centre for the History of Science, Technology and Medicine, University of Manchester.

Eliasson, Gunnar (1991): 'Modelling the Experimentally Organised Economy' *Journal of Economic Behaviour and Organisation*, Vol. 16, Nos. 1-2, pp.153-182.

Eliasson, Gunnar (1996): 'Spillovers, integrated production and the theory of the firm', *Journal of Evolutionary Economics*, Vol.6 No.2, pp.125-140.

Eliasson, Gunnar (1997): 'Competence Blocs and Industrial Policy in the Knowledge Based Economy', *Science, Technology, Industry Review*, OECD Paris 1997-1998.

Eliasson, Gunnar (1998): 'On the Micro Foundations of Economic Growth', in J. Lesourne and A. Orlean (eds), *Advances in Self Organisation and Evolutionary Economics*, Economica: London.

Fai, Felicia and von Tunzelman G. Nick. (forthcoming): 'Industry Specific competencies and converging technological systems: Evidence from patents', *Economics of Innovation and New Technology.*

Forrester, John. W. (1977). 'Growth Cycles', *De Economist*, Vol.125, No.4.

Forrester, John. W. (1981): 'Innovation and Economic Change', *Futures*, Vol.13, no.4. pp.323-331.

Foss, Nicolai (1993): 'Theories of the firm: Contractual and competence perspectives', *Journal of Evolutionary Economics*, pp.127-144.

Franses, Hans Philip (1996): *Periodicity and Stochastic Trends in Economic Time Series*, Oxford: Oxford University Press.

Freeman, Chris (1963): 'The Plastics Industry: A Comparative Study of Research and Innovation', *National Institute Economic Review,* vol.26. pp. 22-61.

Freeman, Christopher (1987): *Technology and Economic performance: Lessons from Japan*, London: Pinter Publishers.

Freeman, Chris (1994): 'Critical Survey. The Economics of Technical Change', *Cambridge Journal of Economics,* Vol.18. pp. 463-514.

Freeman, Chris, Clark, John and Soete, Luc. (1982): 'New Technological Systems: An Alternative Approach to The Clustering of Innovation and The Growth of Industries'. U*nemployment and Technical Innovation.* London: Frances Pinter Publishers. pp. 64-81.

Freeman, Chris, Clark, John and Soete, Luc (1982): *Unemployment and Technical Innovation*, London: Frances Pinter Publishers.

Freeman, Chris and Perez, Carlota (1988): 'Structural Crises of Adjustment: Business Cycles and Investment Behaviour', in Giovanni Dosi; Christopher Freeman; Richard Nelson; Gerald Silverberg and Luc Soete (ed), '*Technical Change and Economic Theory*', London: Pinter Publishers. pp.38-66.

Griliches, Zvi (1957): 'Hybrid Corn: An Exploration in The Economics of Technological Change', *Econometrica*, vol.25, no.4. pp. 501-522

Griliches, Zvi (1990): 'Patent Statistics as Economic Indicators: A Survey', *Journal of Economic Literature*, Vol.28. pp.1661-1707.

Grupp, Hariolf and Schmoch, Ulrich (1992): 'At the Crossroads in Laser Medicine and Polymide Chemistry: Patent Assessment of the Expansion of Knowledge' in Hariolf Grupp: *Dynamics of Science-Based Innovation.* Berlin Heidelberg, New York: Springer Verlag. pp.269-301.

Hart, Peter E. (1971): 'Entropy and Other Measures of Concentration', *Journal of the Royal Statistics Society*, Series A, vol.134. pp.73-85.

Hart, Peter. E (1994): 'Galtonian Regression Across Countries and the Convergence of Productivity', *University of Reading Discussion Paper in Quantitative Economics and Computing*, No.19.

Hodgson, Geoff (1988): *Economics and Institutions, A Manifesto for a Modern Institutional Economics*. Cambridge: Policy Press.

Hughes, Thomas P. (1983): *Networks of Power. Electrification in Western Society, 1830-1930*. Baltimore: Johns Hopkins University Press.

Hughes, Thomas P. (1989): *American Genesis: A Century of Invention and Technological Enthusiasm, 1870-1970*. New York: Viking.

Hughes, Thomas.P. (1992): 'The dynamics of technological change: salients, critical problems and industrial revolutions' in G. Dosi, R. Giannetti and P.A. Toninelli, *Technology and Enterprise in a Historical Perspective,* Clarendon Press, Oxford.

Johnson, Björn (1992): 'Institutional Learning' in Bengt-Åke Lundvall (ed.): *National Systems of Innovation: Towards a Theory of Innovation and Interactive Learning*, London: Pinter Publishers. pp. 23-44.

Kalecki, Michal (1939): *Studies in the Theory of Business Cycles*, [Oxford]: Blackwell, 1966.

Kleinknecht, Alfred (1981): 'Observations on the Schumpeterian Swarming of Innovations', *Futures*, Vol.13, No.4. pp.293-302.

Kleinknecht, Alfred (1987): *Innovation Patterns in Crises and Prosperity: Schumpeter's Long Cycle Reconsidered.* London: Macmillan.

Kleinknecht, Alfred (1990): 'Are there Schumpeterian waves of innovations?', *Cambridge Journal of Economics*, 14, pp.81-92.

Klepper, Steve and Simon, K.L. (1997): 'Technological Extinctions of Industrial Firms: An Inquiry into their Nature and Causes', *Industrial and Corporate Change*, Vol.6. No.2., pp. 379-460.

Kodama, Fumio (1986): "Japanese Innovation in Mechatronics Technology" (*Science and Public Policy*, Vol.13, No.1). Jon Sigurdson (ed.) (1990): *Measuring The Dynamics of Technological Change* (1990). London: Pinter Publishers. pp.39-53

Kodama, Fumio (1992): 'Technological Fusion and the New R&D' in *Harvard Business Review*, No 92407, pp.70-78.

Kondratiev, Nicholas. D. (1979). 'The Long Wave in Economic Life', *Review,* Vol.11 No.4. The original translation can be found in *Review of Economic Statistics*, Vol..XVII No.6. The Russian version was first published in 1925.

Kuhn, Thomas. (1962): *The Structure of Scientific Revolutions*. Chicago: Chicago University Press.

Kuznets, Simon S. (1930a): *Secular Movements in Production and Prices. Their Nature and Their Bearing Upon Cyclical Fluctuations* (Boston: Houghton Mifflin Co., The Riverside Press). 1967 reprint - New York: Augustus M. Kelley Publishers.

Kuznets, Simon S. (1930b): 'Equilibrium Economics and Business Cycle Theory' (*The Quarterly Journal of Economics*, Vol.44. May 1930, p.381-415). *Economic Change. Selected Essays in Business Cycles, National Income and Economic Growth* (1954), Melbourne, London and Toronto: William Heinemann Ltd. pp.3-31.

Kuznets, Simon S. (1940): 'Schumpeter's Business Cycles' (*American Economic Review*, Vol.XXX, No.2, pp.250-271). *Economic Change. Selected Essays in Business Cycles, National Income and Economic Growth* (1954), Melbourne, London and Toronto: William Heinemann Ltd.. pp.105-

124. (This article is a review of Joseph A. Schumpeter (1939) *Business Cycles*, Vol.I and II, New York: McGraw-Hill).

Lindqvist, Svante (1994): 'Changes in The Technological Landscape: The Temporal Dimension in The Growth and Decline of Large Technological Systems', in Ove Granstrand (ed.): *Economics of Technology*. Amsterdam: North Holland. pp.271-288

Lundvall, Bengt-Åke (ed.) (1992): *National Systems of Innovation: Towards a Theory of Innovation and Interactive Learning*, London: Pinter Publishers.

Mahajan, Vijay and Peterson, Robert A. (1985): *Models for Innovation Diffusion. Series: Quantitative Applications in The Social Sciences*. Beverly Hills: Sage Publications Inc.

Mansfield, Edwin (1961): 'Technical Change and The Rate of Imitation', Vol. 29, No.4. pp.741-766

Mansfield, Edwin (1986): 'Patent and Innovation: An Empirical Study', *Management Science*, Vol. 32, No. 2. ,pp.173-181.

Mensch, Gerhard (1975): *Das Technologische Patt: Innovationen Uberwubdeb Die Depression*, Frankfurt, Umschau. English edition (1979), *Stalemate in technology: Innovations Overcome the Depression*. New York: Ballinger.

Merton, Robert K. (1935): 'Fluctuations in the rate of industrial invention' *Quarterly Journal of Economics*, Vol 49. pp.454-474.

Metcalfe, J. Stanley (1981): 'Impulse and Diffusion in The Study of Technical Change', *Futures*, Oct. pp.347-359.

Metcalfe, J. Stanley (1998): *Evolutionary Economics and Creative Destruction*, Routledge: London.

Mill, John Stuart (1964) *Principles of Political Economy* [1862], 2 vols., D. Appleton: New York.

Mowery, David C. and Rosenberg, Nathan (1982): 'Technical Change in the Commercial Aircraft Indistry. 1925-1975'. Nathan Rosenberg (1982): *Inside the Black Box: Technology and Economics*. Cambridge: Cambridge University Press. pp.163-177.

Napolitano, Giovanni and Sirilli, Giorgio (1990): 'The Patent System and The Exploitation of Inventions: Results of a Statistical Survey Conducted in Italy', *Technovation*, Vol.10, No.1. pp.5-16.

Narin, Francis, Noma, Elliot and Perry, Ross (1987): 'Patents as Indicators of Corporate Technological Strength'. *Research Policy*, Vol.16. pp.143-155.

Nelson, Richard R. (1991a): 'Why do Firms Differ, and How Does It Matter?', *Strategic Management Journal*, Vol.12. pp.61-74.

Nelson, Richard R. (1991b): 'The Role of Firm Differences in an Evolutionary Theory of Technical Advance', *Science and Public Policy*, Vol.18, No.6. pp.347-352.

Nelson, Richard R. (1992): 'National Innovation Systems: A Retrospective on a Study', *Industrial and Corporate Change*, Vol.1, No.2. pp.347-374.

Nelson, Richard R. and Winter, Sidney (1977): 'In Search For A Useful Theory of Innovation', *Research Policy*, vol. 6. pp.36-76.

Nelson, Richard R. and Winter, Sidney (1982): *An Evolutionary Theory of Economic Change.* Cambridge, Mass: The Belknap Press of Harvard University.

Noble, David F. (1979) *America by Design: Science, Technology, and the Rise of Corporate Capitalism*, Alfred A. Knopf: New York.

North, Douglas C. (1990). *Institutions, Institutional Change and Economic Performance*, Cambridge: Cambridge University Press.

Ohlman, Herbert (1990): 'Information: Timekeeping, Computing, Telecommunication and Audiovisual Technologies', in Ian McNeil (ed.): *An Encyclopaedia of the History of Technology*, London: Routledge. pp.686-759.

Patel, Pari and Pavitt, Keith (1994): 'The Continuing, Widespread (and Neglected) Importance of Improvements in Mechanical Technologies', *Research Policy*, Vol.23. pp.533-545.

Patel, Pari and Pavitt, Keith (1998): 'The Wide (and Increasing) Spread of Technological Competencies in the World's Largest Firms: A Challenge to Conventional Wisdom', in A. Chandler, P. Hagstrom, and O. Sovell (eds.) *The Dynamic Firm: The Role of Technology, Strategy, Organisation and Regions*, Oxford: Oxford University Press. pp.192-213.

Pavitt, Keith (1984a): 'Patent Statistics as Indicators of Innovative Activities: Possibilities and Problems', *Scientometrics*, Vol.7, No.1-2. pp.77-99.

Pavitt. Keith (1984b): "Sectoral Patterns of Technical Change: Towards a Taxonomy and a Theory". *Research Policy*, Vol.13.

Pavitt, Keith (1988): 'Uses and Abuses of Patent Statistics', in van Raan, *A Handbook of Qualitative Studies of Science and Technology,* Amsterdam: Elsevier, pp. 509-536.

Pavitt, Keith (1986): "Chips and Trajectories: How Does The Semiconductor Influence the Sources and Direction of Technological Change?"in Roy M. MacLeod (ed.): *Technology and The Human Prospect. Essays in Honour of Christopher Freeman*, London and Wolfeboro: Frances Pinter, pp.31-54.

Penrose, Edith T. (1959): *The Theory of The Growth of the Firm,* Oxford: Blackwell.

Perez, Carlota. (1983): 'Structural Change and The Assimilation of New Technologies in The Economic and Social System', *Futures*, Vol. 15, No.5. pp.357-375.

Phillips, Wendy (Forthcoming); 'The Creation of Capabilities and the Emergence of Strategy - the Case of Environmental Regulation and Pilkington', PhD Thesis, PREST, University of Manchester.

Reader, William J. (1975): *Imperial Chemical Industries - A History: Vol. II, The First Quarter-Century, 1926-1952*. Oxford: Oxford University Press.

Reekie, W.D. (1973): 'Patent Data as a Guide to Industrial Activity'. *Research Policy*, Vol.2. pp.246-264.

Richardson, George (1972), 'The Organisation of Industry', *Economic Journal*, Vol.82, pp.883-896.

Rosenberg, Nathan (1976): *Perspectives on Technology*, Cambridge: Cambridge University Press.

Rosenberg, Nathan (1982): *Inside the Black Box: Technology and Economics,* Cambridge: Cambridge University Press.

Rosenberg, Nathan (1994): *Exploring the Black Box. Technology, Economics, and History*. Cambridge: Cambridge University Press.

Rosenberg, Nathan and Frischtak, Claudio (1984). 'Technological Innovation and Long Waves', *Cambridge Journal of Economics*, Vol.8, No.1. pp.7-24.

Rostow, Walt Whitman (1990): *Theories of Economic Growth from David Hume to The Present. With a Perspective on The Next Century*, New York: Oxford University Press.

Sanders, Bev S. (1964): 'Patterns of Commercial Exploration of Patented Inventions by Large and Small Corporations Into Commercial Use'. *Patent, Trademark, and Copyright Journal* (8). pp.51-92.

Say, Jean-Babtiste (1964), *A Treatise on Political Economy* [1834], Augustus M. Kelly, New York.

Scherer, Frederic M. *et al* (1959): *Patents and The Corporation,* Boston, Mass.: privately published.

Scherer, Frederic M. (1983): 'The Propensity to Patent'. *International Journal of Industrial Organization*, No.1. pp.107-128.

Schmookler, Jacob (1950): 'The Interpretation of Patent Statistics'. *Journal of the Patent Office Society*, Vol.XXXII, no.2. pp.123-146

Schmookler, Jacob (1953): 'The Utility of Patent Statistics". *Journal of the Patent Office Society*, vol.XXXV, No.6. pp.407-550.

Schmookler, Jacob (1962): "The Economics of Research and Development. Determinants of Inventive Activity', *American Economic Review*, Vol. LII, No.2. pp.165-176.

Schmookler, Jacob (1966): *Invention and Economic Growth*, Cambridge, Mass.: Harvard University Press.

Schumpeter, Joseph A. (1934): *The Theory of Economic Development: An Inquiry Into Profits, Capital, Credit, Interest and The Business Cycle*, 1967 reprint - London: Oxford.

Schumpeter, Joseph A. (1939): *Business Cycles. A Theoretical, Historical and Statistical Analysis of The Capitalist Process* (Vol. I and II, New York: McGraw-Hill). 1989 reprint (with an introduction by Rendigs Fels) - Philadelphia: Porcupine Press Inc.

Schumpeter, Joseph A. (1942): *Capitalism, Socialism and Democracy*. 1994 reprint - London: Routledge.

Sharp, Margaret (1989): 'Collaboration in the Pharmaceutical Industry - Is it the Way Forward?', DRC Discussion Paper, Science Policy Research Unit, Mimeo, Sussex University, Brighton.

Silverberg, Gerald, Dosi, Giovanni and Orsenigo, Luigi (1988): 'Innovation, Diversity and Diffusion: A Self-Organisation Model', *Economic Journal*, 98, pp.1032-1054.

Soete, Luc L.G. (1987): 'The Impact of Technological Innovation on International Trade Patterns: The Evidence Reconsidered', *Research Policy*, Vol. 16. pp.101-130.

Solomou, Solomos (1986), 'Innovation clusters and Kondratieff long waves in economic growth', *Cambridge Journal of Economics*, 10, pp.101-112.

Sullivan, Richard (1990): 'The Revolution of Ideas: Widespread Patenting and Invention During the English Industrial Revolution', *Journal of Economic History*, Vol. L, No.2. pp.349-363.

Teece, David J. (1992): 'Strategies For Capturing the Financial Benefits From Technological Innovation', Nathan Rosenberg; Ralph Landau and David C. Mowery (eds), *Technology and the Wealth of Nations*, Stanford: Stanford University Press.

Utterback, James (1996), *Mastering the Dynamics of Innovation*, Mass.: Harvard Business School Press.

Utterback, James and Suarez, Fernando (1993): 'Innovation, Competition and Industry Structures', *Research Policy*, 22, pp.1-21

van Duijn, Jacob J. (1981): 'Fluctuations and Innovation over Time', *Futures*, Vol.13, No.4. pp.264-275.

van Duijn, Jacob J. (1983): *The Long Wave in Economic Life*. London: Allen and Unwin.

van Vianen, Ben and van Raan, Anthony F.J. (1992): 'Knowledge Expansion in Applied Science: A Bibliometric Study of Laser Medicine and Polymide Chemistry', Hariolf Grupp: *Dynamics of Science-Based Innovation.* pp.227-267.

Veblen, Thorstein. (1898): 'Why is Economics Not an Evolutionary Science?', *Quarterly Journal of Economics*, Vol. XII. pp.373-397.

Veblen, Thorstein. (1919): *The Place of Science on Modern Civilization.* 1961 reprint - New York: Russell & Russell.

Vernon, Raymond (1966): 'International Investment and International Trade in the Product Cycle', *Quarterly Journal of economics*, 89, pp.190-207.

Vernon, Raymond (1979): 'The Product cycle Hypothesis in a new international environment', *Oxford Bulletin of Economics and Statistics*, 41, pp.255-267.

von Bertalanffy, Ludwig (1968): *General system theory: Foundations, Development, Applications*. London: Allen Lane

von Tunzelmann, Nick. (1995): *Technology and Industrial Progress. The Foundations of Economic Growth,* Aldershot: Edward Elgar.

Walsh, Vivien (1984): 'Invention and Innovation in the Chemical Industry: Demand-pull or Discovery-push', *Research Policy*, Vol. 13. pp.211-234.

Walsh, Vivien (1993): 'Public markets, demand and innovations: the case of biotechnology', *Science and Public Policy*, 20. pp 138-234.

Walsh, Vivien (1997): 'Globalisation of Innovative Capacity in the Chemical and Related Products Industry', in Jeremy Howells and Jonathon Michie (eds), *Technology, Innovation and Competitiveness*. Cheltenham: Edward Elgar.

Watson, Mark W. (1994): 'Business-Cycles Durations and Postwar Stabilization of The US Economy', *American Economic Review*, Vol.84, pp.24-46.

Yakovets, Yuri. V. (1994): 'Scientific and Technological Cycles: Analysis and Forecasting of Technological Cycles and Upheavals', in Ove Granstrand (ed.): *Economics of technology*, Amsterdam: North Holland. pp.397-410.

Index